CAMBRIDGE COUNTY GEOGRAPHIES

SCOTLAND

General Editor: W. Murison, M.A.

T0351977

CAITHNESS AND SUTHERLAND

Cambridge County Geographies

CAITHNESS

AND

SUTHERLAND

by

H. F. CAMPBELL

M.A., B.L., F.R.S.G.S.
Advocate in Aberdeen

With Maps, Diagrams, and Illustrations

CAMBRIDGE
AT THE UNIVERSITY PRESS
1920

CAMBRIDGE UNIVERSITY PRESS
Cambridge, New York, Melbourne, Madrid, Cape Town,
Singapore, São Paulo, Delhi, Mexico City

Cambridge University Press
The Edinburgh Building, Cambridge CB2 8RU, UK

Published in the United States of America by Cambridge University Press, New York

www.cambridge.org
Information on this title: www.cambridge.org/9781107692800

First published 1920
First paperback edition 2013

A catalogue record for this publication is available from the British Library

ISBN 978-1-107-69280-0 Paperback

CONTENTS

CAITHNESS

CONTENTS

SUTHERLAND

ILLUSTRATIONS

CAITHNESS

ILLUSTRATIONS

The illustrations on pp. 5, 6, 9, 17, 24, 28, 38, 43, 63, 64, 65, 66, 76 are from photographs by Mr A. Johnston; those on pp. 26, 120, 122, 148 from photographs by Valentine & Sons, Ltd.; those on pp. 35, 36, 37, 47, 48, 53, 54, 55, 56, 58, 59, 60, 130, 131, 132, 133, 134, 135 are reproduced by permission of the Society of Antiquaries, Scotland; those on pp. 83, 139, 146 from photographs by Mr F. Hardie; that on p. 93 is reproduced by arrangement with Ch. Griffin & Co., Ltd.; those on p. 96. 106, 114, 121, 123, 154 are reproduced by permission of His Grace the Duke of Sutherland; those on pp. 101 and 161 are from photographs by the Geological Survey, Scotland; those on pp. 113, 145, and 158 are from photographs by Mr R. R. Johnstone, Mr D. Leith, and Mr D. Whyte respectively; that on p. 144 is reproduced by permission of Mr J. MacIntosh of Embo.

CAITHNESS

1. County and Shire: Origin and Administration of Caithness

The Anglo-Saxon term *shire* and the Norman-French *county* gradually came into use in Scotland during the twelfth and thirteenth centuries to denote the larger administrative areas. Though Ross and Caithness had been formed into bishoprics in the twelfth century neither district had at that time been constituted a shire. The earliest extant list of the Scottish sheriffdoms is contained in an " Ordonnance " of Edward I in 1305 for the government of Scotland. At this period the sheriffdom of Inverness included the whole of Ross, Sutherland, and Caithness. The country north of the Oykell was in Gaelic named " Cataobh," which may be connected with " Catti," Ptolemy's name for the inhabitants in the second century. Hence the north-east corner of this district got from the Norse the name " Catey-nes."

The first Earl of Caithness of the Sinclair line received in 1455, besides his earldom, a grant of the justiciary and sheriffdom of Caithness. In 1503 the Scots Parliament enacted that, owing to the distance of the northern parts of the sheriffdom from Inverness, courts might be held at Dornoch and " Wik," but the jurisdiction of the sheriff of Inverness was expressly reserved.

Important cases, such as the infeftment or service of the leading nobles, were still taken at Inverness. Sir Robert Gordon's success in getting Sutherland made a shire in 1631 seems to have stimulated the Earl of Caithness to secure similar privileges for his own county. In 1641 Parliament granted a Ratification to the town of Wick, declaring it to be the head burgh of the sheriff-dom of Caithness. The Earl of Caithness was nominated sheriff of the new shire, and towards the end of the seventeenth century this heritable jurisdiction was sold to Sinclair of Ulbster. At the date of the abolition of these jurisdictions in 1747 a sum of £5000 was paid to Sinclair of Ulbster in compensation for the loss of the heritable jurisdiction of Caithness. The sheriff depute (known after 1829 as the sheriff principal) was formerly appointed by the heritable sheriff, but the appointment now belongs to the crown.

In 1853 the counties of Sutherland and Caithness were united into one sheriffdom. In 1870 Caithness was detached from Sutherland and joined to Orkney and Shetland, while Sutherland was united with Ross and Cromarty.

In the administration of public health and education it would seem that ancient relations between Caithness and Sutherland bid fair to be revived, and the two counties were in 1918 united into one parliamentary constituency.

The parishes of the county, nearly all created in the time of Bishop Gilbert (1223-45), have for seven centuries formed the limits of the local ecclesiastical

jurisdictions. The county contains ten parishes con-
stituting the presbytery of Caithness. The two pres-
byteries of Dornoch and Tongue in Sutherland and the
presbytery of Caithness form the Synod of Sutherland
and Caithness, which is co-extensive with the ancient
bishopric of Caithness. The parishes form the areas
for Parish Councils, created in 1894.

Education is administered by the County Education
Authority, created in 1918. The Higher Grade Schools
in Wick and Thurso are practically secondary schools,
while those at Halkirk, Castletown, and Lybster pre-
pare pupils up to the intermediate stage.

Soon after the erection of Caithness into a shire in
1641, Commissioners of the shire were appointed to
provide supplies of men and equipment for the Scots
Army which fought against Charles I. These Commissioners of Supply became the administrative
authority for the county and continued to be so until
the creation of County Councils by the Local Government (Scotland) Act of 1889. The chief executive
officers of the county are the Lord-Lieutenant—head
of the Court of Lieutenancy—and the Sheriff. Under
the Disarming Acts passed in 1716 and 1724 and other
older statutes the Lord-Lieutenant possessed a certain
military jurisdiction, but in 1907 the County Territorial
Force Association took over the military administration
of the territorial forces within the county. Prior to
the war these forces in Caithness were attached to
the Seaforth Highlanders, with headquarters for the
regulars at Fort George and for territorials at Golspie.

The administration of the police of the county is vested in a Joint Committee appointed by the County Council and the Commissioners of Supply. The County Road Board consists of members of the County Council and representatives of the Parish Councils. Public Health is administered by the County Council in the county and by the Town Council in Wick, the only royal burgh in Caithness. Thurso, a burgh of barony, possesses its own burghal administration for general purposes. The administration of state insurance is vested in a County Insurance Committee. Under the Licensing Acts there is a County Licensing Court, composed partly of county councillors and partly of justices of the peace.

In the administration of mental deficiency Caithness is associated with a southern District Board instead of with the neighbouring counties of Inverness, Ross, and Sutherland. For the purposes of agricultural education the county is included in the district of the North of Scotland College of Agriculture.

2. General Characteristics

Stretching in a north-easterly direction between the North Sea and the Pentland Firth, the lowland part of Caithness, wind-swept and treeless, has few scenic features to delight the eye. Yet it has its compensations. It is the same bountiful land, rich in cornfield and meadow, which ages ago attracted the hungry Vikings of barren Norway. The upland moors, the resort of the red grouse, the curlew, and the golden plover, possess the same fascination that yearly attracts thousands of tourists to the Scottish Highlands. These uplands culminate in the striking peaks of Scaraben (2054) and Morven (2313), the latter a familiar landmark to

Fishing Boats leaving Wick Harbour

the fishermen and seamen who frequent the Moray Firth.

The feature of the Caithness coast region is its rock scenery. The Atlantic breakers beating on some of the bolder northern cliffs at a time of heavy gales form an imposing spectacle, and at such times the harbours of Scrabster and Wick experience the terrific forces of nature.

Industrially the Wick herring fishing, though shorn somewhat of its former dimensions, remains the most prominent feature. The outgoing or returning herring fleet is a most attractive spectacle though the newer motor boats and drifters yield in picturesqueness to the old sailing fleet. Caithness has a considerable

John o' Groats

export and import trade, so that the harbour of Wick has for fully a century been a busy centre of industry.

The people of Caithness, whether engaged in seafaring pursuits, in agriculture, or in commerce, have long enjoyed a reputation for sturdy energy and industry. They are upright and straightforward in all their dealings. The county also possesses a goodly number of landed gentlemen interested not only in their private estates, but also in the public welfare.

Many ruined castles to be seen on its rock-bound coasts testify to the turbulent activity of the feudal magnates of Caithness in bygone days.   In recent times no spot in the county is better known than John o' Groats, the Mecca every year of thousands of motorists.

3. Size, Shape, Boundaries, and Surface

Caithness, though only about one-third of the extent of Sutherland, ranks twelfth in area among the counties of Scotland.   Its greatest length extends from the Ord to Duncansby, while the longest stretch from east to west is from Duncansby to Drumhollistan.

The land area of the county (to high-water mark and exclusive of inland water) is 438,833 acres.   Inland lochs are numerous, especially in the parishes of Latheron, Halkirk, and Reay, and cover an area of 7139 acres. The foreshores extend to 2717 acres, while tidal waters cover 131 acres.   Thus the gross area of the county extends to 448,822 acres or 701 square miles.

Caithness forms an isosceles triangle, with its apex at Drumhollistan, the extreme westerly point of the county, just over 30 miles distant from Duncansby and the Ord.   The base of the triangle runs for 40 miles in a north-easterly direction along the Moray Firth, from the Ord to Duncansby, with an outward bulge from Sarclet Head to Noss Head, a gentle inland sweep at Dunbeath and a more pronounced indentation at Sinclair Bay. The north side of the county runs westerly from Duncansby to Drumhollistan, with a northward bulge containing bold and rocky promontories reaching out to Dunnet Head. There is one considerable indentation forming Thurso Bay, with smaller bays at Crosskirk and Sandside. The coastline from the Ord round to Drumhollistan, following indentations, extends to about 90 miles. The inland border of the county touches Sutherland along its whole extent and runs in a southerly direction from Drumhollistan to the Ord, with a westward sweep at Cnoc Gaineimh (1391 feet) about 20 miles south of Drumhollistan, where the water of Berriedale and the Thurso river take their rise.

In Pont's map of Caithness (about 1608), the Halladale is given as the north-western boundary between Sutherland and Caithness, but the watershed has been recognised as the boundary since the erection of the shires. In 1892 Strathhalladale, which, though in Sutherland, formed part of the parish of Reay, was detached from Reay and added to the parish of Farr. A detached part of the parish of Thurso (near Dorrery) was added to Halkirk.

There is a striking contrast between the flat, fertile, and well-cultivated country in the north-eastern angle of Caithness and the inland expanse of barren, waterlogged moor, gradually rising from the northern flats until it attains its greatest elevation in the cone of Morven.

Dunnet Head

The contour of Caithness is dependent upon its geological structure. Where the archæan metamorphic rocks prevail the country consists of barren heather-clad mountains interspersed with many peat mosses, lochs, and bogs, and the scenery closely resembles that of the northern Highlands. The lowland country, on the other hand, in aspect, as in geological structure, resembles the " Laigh " of Moray, the rich haughs of Easter Ross, and the fertile Orkneys. Like the Orkneys, too, Caithness is nearly devoid of trees. In the summer and autumn the rich cornfields, with the pasture and meadow lands, on which splendid herds of cattle comfortably graze, redeem the flatness of the scene. In former times before the days of advanced agriculture, visitors to Caithness were repelled by the unrelieved bareness of the scenery and described the country as " awesome and forbidding."

If Caithness suffers from lack of trees there is compensation in its coast scenery. The deep gèos, the protecting stacks and headlands resound with the crash and roar of huge Atlantic breakers. The elevation of the rocks and promontories adds to the effect. Nowhere in Britain is there finer rock scenery than on the coasts of Caithness.

4. Watershed, Rivers, and Lakes

The ridge extending from the Ord to Drumhollistan, forming the boundary between Caithness and Sutherland, is the watershed dividing the Straths of Halladale and Kildonan in Sutherland from the basins of the Berriedale, the Thurso, and the Forse. Near chalybeate springs on the heights of Knockfin rise the Thurso and the Berriedale, as well as the Halladale. The Thurso (40 miles), known in its earlier course as the Glut or Strathmore Water, flows into Loch More in Halkirk (where it is joined by the Sleiach burn), and thence runs a course of nearly 30 miles into Thurso Bay. The water of Forse rises in Loch Ealach Beg on the confines of Sutherland, one of a number of mountain tarns in close proximity, which discharge themselves in different directions into the Thurso, the Forse, and the Halladale. The only important streams in the county which do not rise at the watershed are the Water of Dunbeath, draining the basin north of the Berriedale, and the Water of Wick, flowing eastwards from Loch Watten, a distance of 14 miles, into Wick Bay and receiving from the southern moors the Acharole and Achairn burns.

The Loch of Watten (N. " Water "), 5 miles long by 1½ miles broad, is the largest lake in the county, and next to it in size are Lochs Calder, Shurrery, and More. There are no fewer than twenty-four lochs in the parish of Halkirk, and a large number also in Latheron and Reay. St John's Loch in Dunnet, like Loch Ma Nathair in Strathnaver, enjoyed in olden times a reputation for its healing virtues. Invalids resorted to it in large numbers, especially at Midsummer (St John's Day), plunging into its waters and going through certain ceremonies which were expected to result in an effective cure.

5. Geology and Soil

The earth's crust is composed of various kinds of rocks, such as granite and syenite, formed by the action of the earth's internal heat ; trap rock and basalt due to extrusion, by volcanic agency, from the earth's interior ; sedimentary rocks laid down in the form of sediment under the waters of lakes and seas, and limestones formed as a rule by masses of minute calcareous skeletons of the lower forms of animal life. The word rock in geology is applied to loose formations of clay, gravel, and even sand. All sedimentary rocks and limestones naturally rest upon a substratum of igneous rock, but in the course of geological time natural disturbances caused much displacement of rock masses. The sedimentary rock discloses a stratified formation which is wanting in the igneous and volcanic rocks. The older sedimentary rocks of the Scottish Highlands have undergone great alterations, caused partly by the earth's internal heat and partly by the pressure of the enormous masses of the later overlying formations. Hence these ancient rocks are called metamorphic.

The following scheme shows :—

THE SUCCESSION OF THE SEDIMENTARY ROCKS
IN CAITHNESS AND SUTHERLAND

Post-Tertiary	1 Recent	The post-glacial peat of Caithness moors : buried forests at Keiss foreshore.
	2 Pleisto-cene	Glacial calcareous drift in north-east of the county (from Moray Firth region) : glacial sandy drift of west and south (from Sutherland).
Pliocene	3 Newer	Fossils consist of marine fauna (wanting in Caithness).
	4 Older	Many of the shells found in these strata are of extinct species (wanting in Caithness).
Miocene	5 Upper	(Wanting in Britain.)
	6 Lower and Oligo-cene	Represented in Isle of Mull leaf bed. The shells are almost wholly of extinct species.
Eocene	7 Upper	Sands and clay (wanting in Caithness).
	8 Middle	Occurs in south of England.
	9 Lower	The London clay.
Cretaceous	10 Upper Chalk and Ware	Surrey white chalk with flints.
	11 Lower Clay and Sands	Weald clay in south of England.

Oolite	12 Upper	Freshwater clay beds between Kintradwell and the Ord : beds above the Brora coal.
	13 Middle	Carbonaceous shales at Brora.
	14 Lower	Shales and calcareous sandstones at Strathsteven, Brora.
Lias	15 Lias	Shales and sandstones with plant remains at Dunrobin : sandstones and limestones in Golspie burn.
Trias	16 Upper 17 Middle	} (Wanting in Britain).
	18 Lower	Red sandstone of Lancashire and Cheshire.
Permian	19 Permian	Magnesian limestone in Yorkshire and Cumberland.
Carbon- iferous	20 Upper	English coal strata.
	21 Lower	Limestones of Lanark and Fife, with shales and ironstone.
Old Red Sandstone	22 Upper	Yellow sandstone of Dunnet—a continuation of the Orkney beds.
	23 Middle	Caithness flagstones at Castletown, etc. (with fossil fish and plants).
	24 Lower	Braemore shales and flagstones with plant remains and conglomerates of Morven.

Silurian and Ordo-vician	25 Upper	The metamorphic rocks of the central highlands and uplands of Caithness.
	26 Lower	The schists of the Caithness uplands underneath peaty moors.
Cambrian	27 Upper	Quartzites in Caithness uplands as at Scaraben : also in west of Sutherland.
	28 Lower	Slates and shales (lower fossils) in Wales.
Archæan Gneiss	29 —	Extend from Cape Wrath to Lochinver, granitic in appearance, but folded and contorted.

The geology of Caithness is rendered somewhat simple by the fact that the surface of the county is made up mainly of two rock formations. The upland parts consist in a large measure of stratified schists, common all over the Scottish Highlands, and Cambrian quartzites. The lowland country, on the other hand, is composed mainly of Old Red Sandstone.

Overlying the metamorphic schists is the sandy glacial drift, covered with peat and dreary moorlands of heather, sedges, and coarse grasses.

Overlying these primeval strata are a series of conglomerates, flagstones, and sandstones in masses which may be said to form the Old Red Sandstone. The same formations occur in Orkney, east of Sutherland, and Ross. The fossil remains, chiefly freshwater fishes, show that the Old Red was formed by deposits in a freshwater lake. In the west of the county geologists

can point to the shore lines of this ancient inland sea.
Sir Archibald Geikie has estimated that the total thick-

ness of the Old Red Sandstone deposits in Caithness and Orkney
exceeds 18,000 feet.   The oldest
beds are the conglomerates formed
by the detrition of the Silurian
schists on which the lake rested.
These conglomerates occur at Berriedale and form some of the higher
mountains in the south-east of the
county.   Morven, Maiden Pap, and
Smean are mainly conglomerates,
while Scaraben, Salvoich, and Scal-
absdale are mainly quartzite.   The
upper beds of the Old Red Sandstones occur in the extreme north
of Caithness, extending into Orkney.
The middle Old Red is exemplified
by the flagstones of Thurso and the
red marls and sandstones of John

**Palæospondylus
gunni**

(Unique fossil fish
from Old Red Flag-
stones, Achanarras
Quarry)

o' Groats district.

In the course of time the older
deposits of the Red Sandstones
were tilted and convoluted by
pressure into mounds and hillocks,
which in turn were worn down by the denuding
forces of rain, frost, and wind.   To these later
periods of the Old Red Sandstone also belongs a sedimentary formation near Dunnet Head, where the

B

Castle Nestaig, Stroma

traces of animal life are chiefly fossil remains of fresh-water fishes.

The soil derived from wearing down the flagstones and sandstones became the subsoil of lowland Caithness. Above it lies the boulder clay of the Glacial Period. The detrition of the flagstones produced a soil containing traces of fertilising ingredients not found in the more ancient gneisses and schists of the interior. The boulder clay of the north-east, formed from ice drifts off the Moray Firth basin, possesses considerable elements of lime, shells, and other organic remains contributing to the fertility of the soil, with the result that north-eastern Caithness became a fruitful land in striking contrast with the barren uplands.

The ice movements during the Glacial Period brought masses of gravel and fragments of rock from the interior into the western and south-western regions of the county, now forming the ridges and eminences of upland Caithness. These boulders and the gravel in which they lie are obviously derived from rocks different from those on which they rest. The drifting masses carried by the ice-flow eroded hollows out of the soil. These hollows in many cases became the beds of lakes or formed marshes, which in time became peat mosses.

In the vast period that has elapsed since the flag-stones of Caithness were laid down the earth's crust has been subject to upheaval and subsidence, greatly altering the surface of the country. The raised beaches so familiar in the east of Sutherland scarcely occur in Caithness. The remains of trees and plants in the

peat mosses furnish ample evidence that Caithness was
at one time covered with forests of pine, birch, alder,
and willow, and that the climate, as well as the elevation of the surface, has varied greatly during the prolonged periods of geological time.

6. Natural History

Occupying the extreme northern limit of the mainland, Caithness possesses certain peculiar features in
its flora and fauna. As in rock structure, so in vege-
table and animal life there is a close connection with
Orkney. At a period subsequent to the Ice Age, when
Britain itself was united with the European continent,
what are now the Orkney Islands formed part of the
mainland. In this post-glacial epoch a sub-arctic flora
and fauna must have gradually overspread the country.
At the present day alpine vegetation is limited in Caithness to the higher mountains such as Morven, Smean,
and the Maiden Pap, with traces upon lower heights
such as Ben Ratha and Ben Nam Bad Mhor.

The alpine vegetation just below the bare windswept
summits of those mountains consists chiefly of mosses
(*Rhacomitrium* and *Arctostaphylos*) mixed with *Cladina*
and other lichens, dwarf whinberries, and stunted
heather. Among the heather will also be found dwarf
shoots of blaeberry (*Vaccinium myrtillus*) and *Alchemilla
alpina*. Patches of alpine bearberry (*Arctostaphylos
uva-ursi*) and other plants beaten down by the wind
form a soft carpet affording a pleasant change to the

hill climber who has been scrambling upon rough screes and boulders.

The lower reaches of the mountains are covered with peat. The Caithness peat hags have been gradually denuded by the action of wind and by drainage, but on the lower slopes of many hills there is still a heavy covering. The prevailing moorland plant is heather (*Calluna vulgaris*), which grows luxuriantly in sheltered places where the conditions are favourable, and shelters a varied undergrowth according to the nature of the soil and the supply of moisture. At one time nearly the whole of Caithness was covered with a layer of heather-clad peat, with an undergrowth of such plants as *Hypnum* and *Hyocomium*, where the drainage was good, and *Scirpus* and *Sphagnums* in the bogs and marshes. Dwarf birches, rowans, alders, and willows grew in sheltered places, while crowberry and *Scilla verna* covered the grassy sward of the lower moors fringing the coast. As the climate became more temperate, and as artificial drainage was introduced, peat gradually receded from a great part of the lowland plains. Heather also gives way in places where the bracken fern flourishes or the whin (*Ulex Europæa*) is common, so that the undergrowth of such plants as *Oxalis*, *Trientalis*, and *Luzula* increases in vigour.

Along the banks of the streams and rivers the vegetation varies according to soil, exposure, the supply of moisture, and the amount of protection afforded by the river banks. Some plants prefer a loam, others a sandy or gravelly bed. Birches, though stunted, occur on the steep banks of various streams with an undergrowth of ferns and mosses. In Langwell and Berriedale valleys the trees attain a goodly growth, but the distribution of scrubwood along the watercourses of Caithness, as a whole, is not so wide as it once was. Hazel is common along some parts of the Dunbeath valley, while the alder and willow are more partial to the Berriedale and the Langwell, perhaps on account of the difference of the underlying rock formation. In the gravelly loams along the river beds there occur *Cynosurus cristatus*, *Festuca*, and *Agrostis*. The loams in the lower reaches of the Thurso and Wick rivers yield *Eleocharis palustris*, *Equisetum*, and *Carex* with sedges, reeds, and rushes.

Along the Reisgill burn there is a varied flora under scrubwood consisting of birch, aspen, hazel, willows, and juniper. The birch woods of the Langwell valley are contracting in area, perhaps owing to deer as well as to the ravages of fungus and the denuding effects of wind. Young birches are not met with so often as seedling rowans. This is explained by the fact that birds help to distribute the berries of the rowan while the seed of the birch is spread chiefly by the action of the wind.

On the rocks and cliffs of the coastal regions are many varieties of lichens and marine algæ. The sandy beaches at Sinclair Bay, Freswick, Dunnet, Thurso, and Reay have sand dunes and hillocks with the usual arenaceous vegetation, such as *Amphilla arenaria*, *Carex*, *and Tortula*, which help to fix the dunes and stretches of sand converting them into "links" as at Keiss and Dunnet. There are reaches of shell-sand at John o' Groat's and Murkle, where the lime gives vigour to the grasses and to attractive flowering plants such as *Primula scotica* and *Parnassia*.

The only important woodland plantations in Caithness are in the Duke of Portland's policies at Langwell, containing not only pine, larch, and other conifers, but a goodly variety of hard woods and ornamental trees.

In the Keiss mounds and primeval rubbish heaps investigated by Samuel Laing traces were found of the reindeer the beaver, the great auk, and the goat. Fragments of the bones of these animals have been found in brochs and in situations indicating that they were contemporary with primeval man. Mr Laing also found traces of the wild ox, the horse, the stag, and the dog. The Norsemen, according to the sagas, hunted the reindeer in the uplands of Caithness.

Among extinct animals which persisted into the historical period were the wolf and the wild boar, which survived in Caithness until the seventeenth century. In the eighteenth century the wild cat, the pole cat, the badger, and the marten were fairly common in the county though now verging upon extinction, if not already extinct. The bat, the hedgehog, and the squirrel are extremely rare and yet they occur in Sutherland. The common seal (*Phoca vitulina*) and the grey seal (*Halichœrus gryphus*) breed in the caves along the coast. Various species of whales occur from time to time in the neighbouring waters, such as the Right whale (*Balæna biscayensis*), the Sperm whale (*Physeter macrocephalus*), and the Rorqual. By far the most common of the cetaceans in British waters are the Finner (*Balænoptera musculus*) and the Sei (*B. borealis*).

The county is rich in bird life. Mr Harvie Brown mentions about 220 different species (including residents and occasional visitors), and there have been numerous additions to the list of Caithness birds since the publication of his book about a quarter of a century ago. Along the coast sea-birds abound in number and variety, while owing to the extension of young plantations at Langwell and Berriedale woodland birds have steadily increased. Among birds now becoming rare are the sea-eagle, the osprey, the raven, and the ptarmigan (sometimes driven from its mountain haunts by injudicious burning of heather). The starling and some species of gull are on the increase.

Of reptiles the adder and the lizard occur in moorland districts, but the slow-worm, common in Sutherland, is somewhat rare. Frogs are plentiful and the toad is common, but the newt is not found.

Fishes abound both in inland and marine waters. Owing to the introduction of mechanical power in sea-fishing many new species of marine fish have become known. Such varieties as witch-soles and megrims, now commercial fishes, were practically unknown before the days of trawling. The industrial shell-fish (lobster, crab, whelk, and mussel) occur in fair abundance, while

Puffins, Dunnet Head

John o' Groat's buckies (*Cyprea Europæa*) well known
to shell collectors, occur on the shores of the Pentland
Firth.

7. Coast Line

The Moray Firth coast line from the Ord to Duncansby
is just over 50 miles ; while from Duncansby to Drumhollistan the curved line is over 40 miles. About five
miles from the Ord is the old Castle of Berriedale, a
noted place of strength in the days of the Norse. Near
the mouth of the Water of Dunbeath stands Dunbeath
Castle, the only one of the older strongholds which is
still inhabited. Below the lands of Forse and Swiney
stands the old castle of Achastleshore not far from
Lybster Bay. Eight miles north-east of Lybster is the
Stack of Ulbster and beyond it Sarclet Head stretching
into the Moray Firth. Four miles farther north stands
the picturesque old Castle of Wick, within a mile of
Wick Bay. Bold cliffs, geos, and havens are a feature
of this coast. The caves provide shelter for numerous
gipsies at all seasons of the year. Masses of Old Red
Sandstone form a bold outcrop on both sides of the
Bay of Wick, where the North Sea breaks in spray and
foam. There are stretches of sandy beach in Sinclair
Bay and Freswick Bay, with sites of ancient strongholds. Castles Sinclair and Girnigoe are associated
with the turbulent Earls of Caithness of the sixteenth
and seventeenth centuries. Ackergill Tower was a
leading stronghold at an earlier date. Keiss and

Brig o' Trams, Wick

Bucholly Castles were also places of strength belonging to feudal magnates of the olden times.

The Pentland shore extends for 15 miles from Duncansby to Dunnet Head. Parts of the coastal region are flat or undulating, but the coast scenery at Duncansby and round the promontory of Dunnet retains its bold and rocky character. Some notable rocks are named the Men of Mey and the Stacks of Ham. The largest indentation on the Caithness coast is Thurso Bay, with a coast line (following the curve) of about 15 miles. Dwarwick Head on the coast of Dunnet and Holborn Head are striking promontories. In the 12 miles of coast between Holborn Head and Drumhollistan are many bold and lofty cliffs shelving into the Atlantic. The most striking promontory is Brims Ness, while, as the name denotes, there is a stretch of sandy beach at Sandside.

8. Coastal Gains and Losses : Lighthouses

It has been stated that in the early Pleistocene period Orkney and Caithness were united. Owing to extensive subsidence of the land in the glacial or post-glacial period the lowland valleys of Orkney were submerged and what had been a peninsula was converted into a group of islands. The low-lying country between Hoy and Caithness became a maritime channel, now known as the Pentland Firth, while an extensive area to the north-east of the Firth was submerged.

Tinkers' Cave

There was no subsequent upheaval comparable with this primeval subsidence, and raised beaches form no prominent feature of the Caithness coast. In Sinclair Bay traces are to be found of submerged forests extending under the North Sea, indicating subsidence of the land in the post-glacial epoch. Stroma and the Pentland Skerries were the higher hilltops in the country submerged by the great subsidence.

Unlike those of Sutherland, the coasts of Caithness are free from shoals and dangerous reefs, but the Pentland currents have always been a source of danger to shipping. In olden times the " bores of Duncansby " and the " Men of Mey " were scenes of many a shipwreck. A lighthouse was erected upon Pentland Skerry in 1794, to guide the navigation of the eastern entrance to the channel. The lofty Dunnet Head light was provided for tne western entrance in 1832. Noss Head light, near Wick, was erected in 1869, and Holborn Head light, near Thurso, in 1862. No light has been erected upon Duncansby, though it forms the northeastern projection of Scotland, but a light has been provided on the northern projection of Stroma known as the Redhead, which, according to some authorities, is the " Verrubium " of Ptolemy.

9. Climate and Weather

The atmospheric conditions prevailing in Caithness depend as a rule upon two widely-separated centres of pressure. To the south of Iceland there is an oceanic centre of atmospheric pressure round which sweep cyclonic disturbances affecting north-western Europe, chiefly during autumn and winter. In spring and summer the centre of disturbance moves to the eastern regions of European Russia, round which there are low pressure or cyclonic conditions in summer and high pressure or anti-cyclonic conditions in autumn and winter. In winter the atmospheric conditions in the regions surrounding these two centres of pressure help to give the atmosphere a movement in the same direction, that is from west to east or north-east, thus causing strong westerly gales (often accompanied with rain or sleet) which sweep over Caithness with a keenness, partly the cause and partly the consequence of its treeless condition. These gales raise the huge Atlantic billows which lash the cliffs on its western and northern shores.

During anti-cyclonic movements the weather is usually calm and fine, and in winter accompanied by keen frost. A feature of the spring weather is the prevalence of cold, chilly, easterly winds, blowing from continental Russia and Scandinavia. These winds are sometimes accompanied by sea fogs noted for their depressing effects. When these easterly winds rise to a gale the surface of the North Sea becomes greatly disturbed and breaks with terrific force upon the eastern coasts. Time and again the breakwater of Wick has been destroyed by violent easterly gales and all the efforts of man have hitherto been unable to meet the onset of the breakers. Easterly winds are common up to about the time of the summer solstice, when Caithness summer weather really begins. Autumn sets in about the middle of August.

The rainiest months of the year are January, November, and December, the average monthly rainfall then being about 6 inches compared with an average for Scotland of 3.73 inches in January, 4.77 in November, and 3.91 in December.

The driest months are usually July and August, when the average rainfall is from 1½ to 2 inches for each month, which is rather above the Scottish average. On an average rain falls upon twenty-two days in each of the three winter months and twelve days in each of the months of July and August. The total average rainfall in Caithness during 1913 amounted to 51.79 inches compared with an average for Scotland of 39.86 inches. Rain fell in Caithness upon 225 days throughout that year, the Scottish average being 201 rainy days.

The mean temperature for the three winter months is 39.5° Fahr. compared with a mean of 39.1° for Scotland. The highest temperature is attained in June, July, and August, when there is a mean of 55° compared with 56.2° for Scotland.

In June 1913 there were 170 hours of sunshine in Caithness, compared with an average of 169 hours throughout Scotland. On the other hand the average sunshine in the county in December 1913 extended over 21.7 hours, compared with 26 hours for the whole of Scotland.

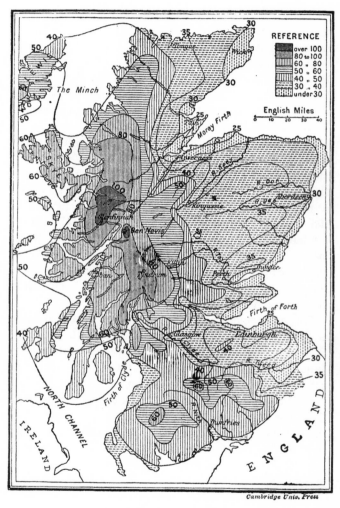

Rainfall Map of Scotland
(*By Andrew Watt, M.A.*)

10. The People—Race, Language, Population

Long before Caithness became the meeting place of Norsemen and Gael its inhabitants had become a composite race. The researches of Laing and others prove that the aboriginal inhabitants of northern Scotland belonged to a long-skulled race, who used weapons of stone and lived on fruits and fish. The first immigrants into Britain were a dark-haired race whose weapons were of smooth stone (neolithic, *i.e.* new stone) and who lived chiefly by rude tillage and hunting. The neolithic people occupied the country many centuries before the historic period, and indeed may be referred to a time when Britain formed part of the European continent. The neolithic races may have spread northward from the Mediterranean region until they reached the most northern parts of Britain. Some writers claim that traces of a palæolithic (*i.e.* old stone) people have been found in Scotland. This is not gener ally admitted, yet the oldest remains found in Caithness may, according to Dr Munro, belong to a pre-neolithic people. Conquered by successive hordes of invaders but never exterminated, the neolithic race intermingled with its conquerors, producing a composite race of descendants, who again combined with the successive waves of conquering immigrants to produce the composite people of to-day. In the existing population of Caithness there remains a racial admixture of the

c

neolithic people and of the successive immigrations into northern Scotland.

In the period following the separation of Britain from the continent, there occurred an immigration into Britain from Central Europe of a short-skulled people, who imposed their language and customs upon the neolithic races. While they coalesced with these races, yet to a certain degree they must have remained a ruling or aristocratic caste. In time this short-skulled race spread over the whole of Britain, penetrating to the shores of the Pentland Firth and the islands beyond Cape Orcas. Besides having tools of flint, they used bronze in the manufacture of their weapons. Though the art of smelting iron was unknown to them, they were able to work copper and tin, which they wrought into bronze, and so the period of their civilization is named the Bronze Age. The next immigrants were the Brythons from Gaul, who introduced the use of iron and Celtic art, and who possessed the country when the Romans came. At first called Britons by the Romans, the northern Britons were styled Caledonians by Tacitus, but from the third century were known to the Romans and the southern Britons as the Picts.

The next invasion came from the west. The Scots—a Gaelic race—crossed from Ireland in the Roman period, subdued Argyll and the Western Isles, and encroached gradually upon the Pictish kingdoms. The Gaelic invasion had in the sixth century developed a missionary movement led by Columba, whose followers soon carried the Gaelic language and Gaelic civilization to the remotest corners of the land. Many of the Columban missionaries became identified with northern localities. Dunnet was associated with St Donat, whose chapel is the " fanum Donati " of Buchanan. The patron saint of Wick was St Fergus, who is said

Bronze Armlet

to have flourished in the eighth century. It would seem that in the course of the seventh and eighth centuries Gaelic, introduced by the Christian missionaries, rapidly supplanted Pictish (which possessed no written literature), just as in later centuries Gaelic supplanted Norse in the Hebrides.

The Norse began to find their way into Caithness in the ninth century. Norse jarls soon intermarried with the Gaels and settled on the land. Owing to the superior influence of their sea power and the remoteness of Caithness from the seat of Scottish government, the north-eastern part of the county soon fell into the possession of the Norse, who gradually spread over the fertile lowland plains and became exporters of grain

Norse Brooch of Bronze
(From Yarhouse)

and cattle to barren Norway. The Gaels gradually receded to the less fertile uplands of Latheron, Halkirk, and Reay, while retaining a hold upon parts of Watten and Bower. The partition of the country between Norseman and Gael thus effected in the tenth century remained undisturbed for nearly a thousand years. The place names of the county bear ample testimony to this. Most of the names of hills, streams, lochs, and townships in the upland parishes are of Gaelic origin, a few being Pictish, while the names in the north-eastern or lowland parishes are mainly Norse or English, with a sprinkling of names of Gaelic or Pictish origin.

The population of 32,010 (according to the census

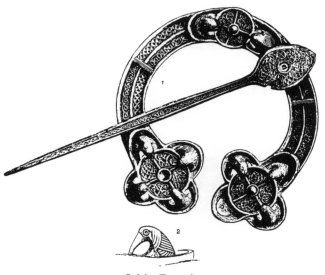

Celtic Brooch

of 1911) included 3251 persons who were not natives of the county. Nearly every other county in Scotland contributed a quota of these, but the larger number of them came from the neighbouring counties. The Gaelic-speaking population in 1911 was 1680, mostly in Latheron and Reay, and a smaller number in Halkirk,

Thurso, and Watten. Gaelic does not appear to receive any encouragement in the schools and is not largely spoken by the young. The population of Caithness, as of other northern counties, has been steadily declining during the past half century. The Burgh of Wick shows some signs of expansion, but Thurso is practi-

Fish-workers

cally stationary, while the rural parishes show a steady decline of population at each successive census. The density of the population in Caithness is 46 to the square mile compared with 157 to the square mile for Scotland. Wick contains nearly one-third of the whole population of the county. From rural Caithness, which is almost wholly agricultural, large numbers belonging to the farming community have been emigrating in recent years, chiefly to Canada.

According to the census of 1911 there were employed in the county 494 coopers, 140 millwrights, blacksmiths, and strikers, 73 men fish workers and 196 women fish workers. Carpenters, painters, plumbers, and masons numbered 494, tailors and shoemakers 247, bakers and grocers 268, and general labourers 248. There were 302 women employed as dressmakers. The number of fishermen was 637 while there were 189 fishermen-crofters.

11. Agriculture

Caithness like Orkney became wealthy in the Norse period owing to its corn trade. So great was the agricultural wealth of Orkney and Caithness at that time that their jarl took precedence at the court of Norway over all the other Norse nobility.

For centuries there was an export trade in cattle and corn from Thurso to the lowlands of Scotland and to the Baltic ports, yet the methods of cultivation remained primitive till towards the end of the eighteenth century. The system of short leases and payment of rent by services as well as in kind was detrimental to progress in agriculture. In the eighteenth century, however, an era of improvement was inaugurated by Sir John Sinclair, who about 1770 took the lead in forming the Board of Argiculture—then a sort of voluntary organization under government patronage. By the end of the century enclosure of fields, rotation of crops, and improvements of stock and implements became general among the landlords of the county. Potatoes were introduced about 1760 and the cultivation of a turnip crop became general after 1780. Large farms were formed by throwing together a number of small holdings. Messrs Paterson and Purves deserve mention as great improvers of land and stock in the early part of the nineteenth century. Large grazing farms were formed in the upland parts of the county. Cheviot and black-faced sheep replaced the old native breeds. The dwellings of the smaller tenants as well as their offices were greatly improved.

Towards the end of the nineteenth century the county acquired a good reputation for high-bred horses and cattle. The Clydesdales, shorthorns, and Border Leicesters of Caithness rank among the best. The reputation for cattle and corn is now such as enables Caithness to take a leading place among the agricultural counties of Scotland.

In 1913 there were 5652 horses in the county, of which 4074 were in use for the purposes of agriculture. There were 21,539 cattle, of which 6420 were cows in milk. The total number of sheep was 126,775 and of pigs 1292.

Out of a total area of 438,833 acres there were 111,617 under crops and grass. The chief crop is oats, which in 1911 was grown upon an area of 31,512 acres, producing 145,849 quarters of grain—a quantity exceeded in none of the other Highland counties and in only eight counties of Scotland. No wheat is grown in the county as the surface is too cold and bare. In 1911 rye was grown only upon 4 acres, peas 3 acres, and onions 1 acre. The area under potatoes in 1915 was 1498 acres. The potato crop is usually good. Hay and pasture, which covered 36,954 acres in that year, do not as a rule give a large return.

There are 68 farms in Caithness of over 300 acres and this number has remained constant for a good many years. On the other hand, during the past twenty years the number of holdings above 50 acres and under 300 acres has increased from 360 to 397. In 1911 the number of holdings under 50 acres which were rented from proprietors was 2231, while only 28 small holdings were in the occupation of the owner. Of the holdings above 50 acres, 18 were occupied by the owner and 447 rented or mainly rented from proprietors.

According to the hypothetical estimates of the value of produce contained in the Agricultural Statistics for 1913, the produce of Caithness for that year was as follows :—

Barley .	.	3,570 qrs. at 29/2	.	.	£5,206
Oats .	.	145,849 ,, ,, 20/10	.	.	151,926
Potatoes	.	7,585 tons ,, 54/–	.	.	20,479
Hay .	.	10,951 ,, ,, 75/6	.	.	41,340
				Add	
Value of horses .	.	.	.	.	.	150,000
,, cattle .	.	.	.	.	.	340,000
,, sheep .	.	.	.	.	.	200,000
,, pigs and poultry .	.	.	.	.	10,000

£918,951

12. Fishing and other Industries

Like Lerwick, Leith, and the Aberdeenshire ports, Wick is a leading centre of the Scottish fishing industry. When the harbour is extended an repaired in accordance with present plans, the position of Wick will no doubt be improved. The proposed extension is estimated to cost £150,000. Grants in aid are to be given by the Development Commissioners and by the Treasury.

The quantity of fish landed at Wick in 1911 was 553,868 cwts., valued at £185,106. The quantity for 1911 had been exceeded in the previous year, but the value for that year constituted a record. During the herring season in July and August many boats from other ports fish at Wick. In 1911 the fishing fleet (including both home fleet and strangers) consisted of about 200 steam fishing boats and 240 motor and sail boats, of which 23 steam vessels and 55 motor and sail boats are registered at the port of Wick. The value of fish landed at Keiss and Lybster in 1911 was £1460, mostly obtained for shell fish (lobsters and crabs). The value of the Stroma fishing for 1911 (chiefly cod and lobster) amounted to £1550, while that at Thurso and Scrabster yielded £2099. The number of local men engaged in fishing is everywhere declining. It would seem that the younger men are not taking to a seafaring life and the inshore white fishing is declining at several of the smaller stations. Prior to the war the foreign shipments of herring to Germany and Russia

Wick Harbour

were considerable, amounting in 1911 to over 160,000 barrels. There is also a growing trade with America.

Wick has several subsidiary industries, such as barrel making, net making, boat building, and others which flourish alongside of the fishing industry.

The salmon fishings of Thurso river are of considerable importance, the assessed rental for 1911 being £1638. District Fishery Boards have recently been constituted for the waters of Wick and Dunbeath. In many of the lochs and smaller streams good baskets of trout are to be obtained, and these fishings form an attractive recreation to numerous sportsmen.

The flagstone industry, which was founded at Castletown and Thurso in 1837 by Mr Traill of Rattar, formed a valuable source of wealth to the county but, owing to the introduction of concrete pavements and manufactured paving blocks, this industry has declined.

Caithness is not a mining county. To meet local needs, good building stone is quarried in various parts of the county. The older rock formations provide harder and coarser material employed in the manufacture of road metal. Limestone is scarce and a good deal has to be imported for agricultural purposes. The number of persons engaged in mines and quarries in 1911 was 114 and there were 62 stonecutters.

13. Shipping and Trade

Until the rise of the herring-fishing industry at Wick towards the end of the eighteenth century, Thurso was the chief centre of trade and shipping in Caithness. Prior to that time Wick was a mere village, while Thurso maintained a considerable trade, not only with the south of Scotland but with the Low Countries and the Hanse towns.   Indeed it continued for centuries to occupy a position among the Scottish seaports.

The importance of Wick as a trading port began in 1808, when the harbour was constructed and Pulteney-town founded by the British Fishery Society.   From that time there was a rapid extension in the export trade in herrings, and the population of Wick soon exceeded that of Thurso.   About 1840 regular steam communication was established between Wick and Aberdeen.

From that year until the extension of the railway to Wick and Thurso in 1874, the steamers of the Orkney and Shetland Steam Navigation Company provided the chief means of transport of Caithness produce to the southern markets as well as of import of necessary commodities.   There soon followed a reduction in the number and tonnage of coasting vessels, which had for many generations maintained the trade of Caithness with the outer world.   The extension of the railway was a further blow to this coasting trade, which has now shrunk to a few vessels in the coal trade.   The only other harbour of any importance on the Caithness coast is the fishing port of Lybster.

14. History of the County

In the earliest period of which records exist Caithness was occupied by the Picts. As late as the ninth century the adjoining firth got from the Norse the name of the " Petland " (Pictland) Firth. The spread of Gaelic culture and institutions followed upon the introduction of Christianity by the disciples of St Columba in the seventh century. The Christian missionaries affected language as well as religion, for the Pictish language was gradually supplanted by Gaelic. The Norse on their arrival in the ninth century found the country ruled by maormers administering customary laws and jurisdictions.

In the tenth and eleventh centuries Norsemen gradually settled in lowland Caithness. Along with Orkney Caithness became a valuable source of food supply for Norway, which in its whole extent contained barely as much fertile and level land as there is in Caithness. Skilful seamen and capable shipbuilders, the Norse were well acquainted practically with the influence of sea power. Occupying a remote situation and separated by the Highlands from central Scotland, Caithness fell an easy prey to Norse vikings. The Jarl of Orkney, head of the Norse power in the British Isles, soon acquired control over Caithness and the east coast of what they called " Sudrland " (Southland). He then assumed the title of Jarl of Orkney and Caithness.

The Gaelic-speaking inhabitants meanwhile withdrew

Ogham Stone, Latheron

to the upland districts, where Celtic chiefs for centuries

Cross with Runic
Inscription, Thurso

administered the customary Celtic laws. When the Norse were converted to Christianity in the eleventh century, their clergy were not at first brought fully into the organization of the Roman Church. The metropolitans of York and Hamburg contended for ecclesiastical jurisdiction over the Norse churches. About 1150 the country beyond the Dornoch Firth was formed into a diocese by David I of Scotland. The first bishop of the new diocese, named Andrew, had been a canon of Scone and enjoyed some reputation for learning. He had his "High Church" at Halkirk (High Kirk) on the borders of the Norse and Celtic parts of his diocese. In 1198 his successor, Bishop John, was savagely attacked by a follower of Jarl Harold and subjected to barbarous mutilation. Such treatment of a church dignitary aroused the indignation of the country. William the Lyon invaded Caithness in 1202 and compelled the Norse Jarl Harold Maddadson to make submission and to deliver up his daughter Matilda as a hostage for his good behaviour. It is doubtful whether the kings of Norway claimed any jurisdiction on the mainland after this incident, though King Hacon levied tribute upon Caithness in 1263. From the beginning of the thirteenth century the jurisdiction of the Jarls of Orkney and Caithness was limited to the north-eastern or lowland part of the diocese.

About 1220 another bishop of Caithness was subjected to outrage. Owing to a dispute about church dues Bishop Adam was seized by the people and burned to death in his own kitchen at Halkirk. In consequence of so serious an offence, a punitive expedition to Caithness was undertaken by Alexander II. Jarl John Haraldson, who had been party to the crime, was severely punished, and it is said that the Scottish king cut off the hands and legs of eighty men concerned in the murder of the bishop. A man of great administrative capacity, equally noted for his piety and his culture, Gilbert, Archdeacon of Moray, was appointed bishop of Caithness. Under him (1223-45) the cathedral chapter was constituted and the whole diocese divided into parishes. The parish endowments of Sutherland and Caithness were allocated to the various dignitaries and canons. The prebend of Thurso, Halkirk, and some other parishes was assigned to the bishops, who enjoyed these endowments until the abolition of episcopacy, when they lapsed to the crown, which in consequence owns valuable domains in Caithness to this day.

On the death of Jarl John Haraldson, in 1231, without an heir, the direct line of the Norse Jarls became ex-

D

tinct. The Scottish king conferred the earldom of Caithness upon Magnus, son of the Earl of Angus and probably son or husband of Matilda, daughter of Jarl Harald. Magnus was at the same time Jarl of Orkney under the king of Norway, and Earl of Caithness under the king of Scotland. The Angus line of the earls of Caithness became extinct in 1329 and the earldom was then conferred, first on a branch of the Stewarts, and thereafter on the Crichton family. The Crichtons became extinct about the middle of the fifteenth century.

In 1455 James II granted the earldom of Caithness to William Sinclair, second son of the Jarl of Orkney, who was chancellor of Scotland and one of the leading Scottish statesmen of the period. The earldom has continued in the Sinclair family to the present day, although the estates no longer go with the title. The present Earl of Caithness owns estates in Aberdeenshire.

The advent of the Sinclairs marked a period of much unrest and strife in the county due to feudal rivalries.

After the death of Alexander III Scotland continued for centuries in the grip of feudalism, and owing to the weakness of the government little progress was made by the people, who were kept in continual turmoil by the rivalries of local magnates. In 1426 there was a skirmish at Harpsdale between the Gunns (a Norse clan) and the Mackays of Strathnaver. Again in 1438 the Gunns and the Keiths fought on the moor of Tannach, near Wick. These clan contests continued until near the end of the seventeenth century. In 1667 the Mackays raided Caithness. In the following year the Sinclairs raided Strathnaver and took a "creach" of nine hundred cattle. In 1680 Lord Glenorchy, who had purchased the lands and title of the earldom from George, Earl of Caithness, invaded Caithness with an army of 700 Campbells to make good his claims. He was opposed by Sinclair of Keiss, who claimed to be the next heir to the earldom, and a fight took place at Altimarlach, two miles west of Wick, where the Sinclairs, partly by stratagem, were completely defeated. This was the last clan battle fought in Scotland. Glenorchy's piper, Findlay MacIver, composed at this time the well-known marching tune, "The Campbells are Coming." The government eventually recognized the claim of George Sinclair to the earldom, and Glenorchy, by way of recompense, received the title of Earl of Breadalbane. Glenorchy had found it impossible to collect his rents in Caithness. The people regarded him as an intruder and a usurper. The lands of the earldom passed into the possession of the Sinclairs of Ulbster, who also acquired the heritable jurisdiction. Some years afterwards Sir John Sinclair of Ulbster entered upon those improvements of the agriculture of the county with which the modern period of industrial progress began. Other landowners followed suit and there was inaugurated an era of progress which has continued to the present day.

15. Antiquities

Caithness is rich in antiquities and these have been studied by antiquaries of eminence such as Cordiner, Rhind, Laing, Barry, and Curle. No part of the literature of the county is so full and so varied as its antiquarian lore.

According to Dr Munro the shell and rubbish heaps excavated by Laing in 1866 belong to the period of transition from the palæolithic to the neolithic age. If this is correct they are among the oldest traces in Scotland of human civilization. Among the other monuments of greatest antiquity in Caithness are the long, horned sepulchral cairns belonging to the neolithic period, usually containing chambers of one or more compartments, in which are found human remains, flint implements, rude ornaments, and pottery.

The circular cairns throughout the county are of later date, but they also yield relics of the neolithic period. The pottery is more elaborate in design while the implements also indicate a more advanced people. Articles of bronze are now to be found as well as flints. In these cairns the burial cists are smaller than those of the long cairns and have no access from without.

The stone circles found all over England and Scotland belong to the Bronze Age. Burial cists frequently occur within these circles, and some authorities believe that they . were erected for religious purposes at a

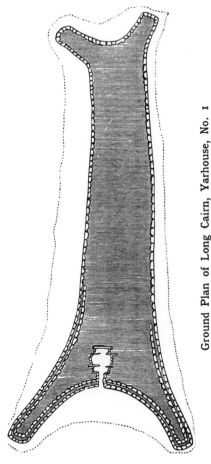

Ground Plan of Long Cairn, Yarhouse, No. I

(240 feet in length)

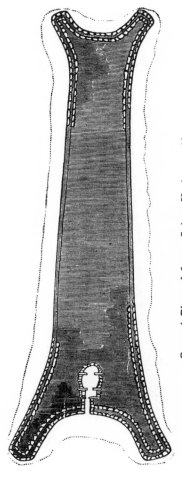

Ground Plan of Long Cairn, Yarhouse, No. 2

(190 feet in length)

time when forms of ancestor worship prevailed. Stone circles, which are numerous in Sutherland, were at one time more common in Caithness, but they have nearly all been swept away. In a number of cases only a single

View of Passage and Chamber, Long Cairn,
Yarhouse, No. 1

boulder or stone now remains where at one time there was probably a stone circle. More frequently in Caithness rows of stones or boulders are to be found which are also referable to the Bronze Age and associated with burial cairns or cists. Some of the standing stones in the county are of great size, such as " Stone Lud " at Bower.

These huge stones sometimes occur in pairs, as at Yarrows

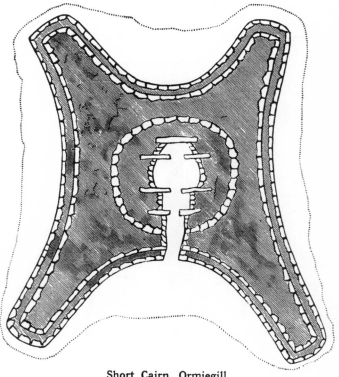

Short Cairn, Ormiegill
(66 feet in length)

and Achvarasdal. Stones of later date mark the burial place of some noted Celtic maormer or Norse warrior.

Of the dwellings of the earlier neolithic race no

traces remain. The dwellings of the Bronze Period were circular, and " hut circles," though not so numerous in Caithness as in Sutherland, occur at Greenland in Dunnet, at Warehouse, and at various places in Latheron, such as Forse, Langwell and Berriedale.

After the Bronze Period came what is called the Iron Age, associated with the advent of the Celtic people who over-ran Britain some five centuries B.C. The Celts of the Iron Age had simpler burial customs than the neolithic races. They placed their burial cists in the earth without any cairn or other permanent mark. These cists, usually small, contain along with human remains specimens of the Celtic art of the period. The chief monuments to this people are their fortified dwellings or brochs. There are remains or evidences of 145 brochs in Caithness, most of them now little more than green mounds. The late Sir F. T. Barry of Keiss excavated twenty-four brochs in different parts of the county and was thus enabled to furnish detailed plans and measurements of these structures. The brochs served both as dwellings and as places of strength. Roman relics, chiefly glass and pottery, have been found in some of the brochs, a fact which would indicate that they were in use in the earliest centuries of our era. Near some of the brochs are hill forts or fortified enclosures, such as occur at Holborn Head, and at Loch Watenan, near Wick. Castle Linglas, on the shores of Sinclair Bay, at one time supposed to be an old Norse stronghold, is really a broch. The excavations of some of the brochs yielded querns, weaving combs

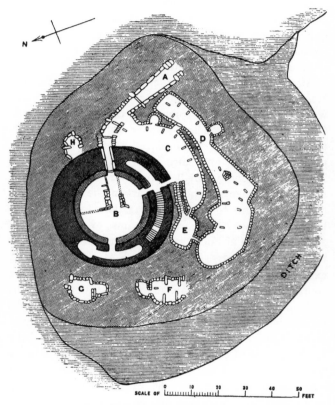

Ground Plan of Broch, Yarhouse

and spindles, as well as articles of personal adornment. These structures are referable to a period which must have extended over many centuries. The latest of them might have been in use when the Gaelic missionaries reached Caithness. The builders of these brochs were people of relatively advanced culture, possessing

a good practical knowledge of farming and high skill in masonry. They knew how to choose the best cultivable land in the county for their settlement, while the skilful design of their brochs has been much admired.

The earth houses, of which examples occur at Langwell in Latheron and Ham in Dunnet, belong to the Iron Age. Probably owing to the obdurate quality of the subsoil these structures are not so numerous in Caithness as

Silver Tankard from a " Picts House " in Canisbay

in Sutherland, but several examples survive of galleried dwellings.

The early Christian monuments consist chiefly of sculptured stones, sometimes rudely inscribed with ogham script and having interlaced Celtic ornamentation. The stones with carved symbols are probably of pre-Christian date. The Skinnet Stone, whose fragments are preserved in Thurso Museum, and the Ulbster Stone, now in the grounds of Thurso Castle (brought thither from St Martin's Church at Clyth), have beautifully incised Celtic crosses on both sides. Sculptured stones with various symbols and ogham characters were found at Keiss some years ago, and are now in the National Museum at Edinburgh. An incised stone with a runic inscription, also in Thurso Museum, is to be assigned to the Norse Christian period, which dates from the death of Sigurd the Stout in 1014.

Tortoise Brooch from Castletown
(Characteristic Ornament of the Viking Period)

Of the Norse Pagan Period prior to that date few relics survive. Two brooches, a bracelet, and a pin, all of the Pagan Period, were found in a burial cist at Castletown in 1786. Brooches and bracelets were also found near Wick about 1837, and at other places in the county at subsequent dates. Two interesting sculptured stones with incised symbols are preserved at Sandside House.

The hill forts on Ben Freacadain and Cnoc an Ratha in Reay, and at Garrywhin in the parish of Wick are probably to be referred to the Iron Age.

16. Architecture—(a) Ecclesiastical

Nothing now remains of the many churches or cells of the Columban Church which existed throughout Caithness in the seventh, eighth and ninth centuries. According to a recent writer the location of the sites of about thirty can still be indicated. The architecture of these primitive churches was of the simplest character. Having walls of turf or dry built flagstone, they were roofed with " divot " and were entirely devoid of architectural ornament. In the troublous times of the pagan Norse occupation most of these churches must have been destroyed.

With the conversion of the Norse there began a new period of church architecture in Caithness, but the Norse churches of the eleventh and twelfth centuries have almost entirely disappeared. The chancelled ruins of St Mary's at Lybster (Reay), St Thomas's at Skinnet, and Gavin's Kirk at Dorrery retain some features which may be referable to this early Norse period. St Mary's Church, which is the most interesting ecclesiastical ruin in the county, closely resembles in its main features the Church of St Peter in the Orkney island of Weir. It has no windows and the jambs of the doorway slope inwards.

When Gilbert succeeded to the bishopric of Caithness in 1223, the Early English style of Gothic architecture, already prevalent in the south of Scotland, extended its influence to Caithness. Gilbert erected his cathedral at a safe distance from the turbulent Norsemen, but did not neglect the northern division of his diocese. The oldest portions of the Church of St Peter's at Thurso appear to belong to this period, though the greater part of the existing ruins are referable to the sixteenth century or even later. The window in the south transept of St Peter's, with its five lights, resembles the west window in the nave of Dornoch cathedral. St Peter's Church, the most imposing ecclesiastical ruin in Caithness, deserves to be protected from further decay.

One or two of the parish churches possess features of some antiquity. The churches of Dunnet and Canisbay, with their big square towers (with saddle-back roofs) probably belong to the fourteenth century. Canisbay church, however, was considerably altered at various dates and especially in the eighteenth century. The old church of Reay was reconstructed and used as a place of burial by the Mackays of Bighouse. The chapel of St Magnus, which formed part of the Hospital, is probably to be assigned to the Norse period. The bell tower or belfry of Latheron, situated on a hill about 500 yards from the church, is a somewhat unique architectural feature. It was erected about the end of the seventeenth century.

17. Architecture—(*b*) Military, Municipal and Domestic

The Norse have left but scanty remains of a military character, their power having been upon the sea. It is said that Castle Linglas was used as a stronghold by Sweyn Asleifson.   The period of Roman Catholic ascendancy in the twelfth and thirteenth centuries was on the whole a period of civil peace, and no military

Ackergill Tower

monuments belonging to that period now survive. The keeps and feudal strongholds of Caithness are to be assigned to the period following the War of Independence, when the central government of Scotland was weak and the local barons turbulent and quarrelsome.   These strongholds were usually erected on rock-bound pro-montories on the eastern and northern seaboard.   Such were Berriedale Castle, The Old Man of Wick, Forse Castle and Buchollie Castle. Girnigoe Castle and the Tower of Ackergill date from the fifteenth century, while Barogill Castle, formerly the seat of the Earl of Caithness, and the picturesque Castle of Keiss (also an appanage of the Sinclairs), were erected towards the end

Castles Sinclair and Girnigoe

of the sixteenth century. The castles at Old Wick, Forse, and Keiss were merely towers or keeps, while those at Buchollie, Girnigoe, and Barogill were on a more elaborate plan, possessing courtyards with ample accommodation for the feudal magnate and his retainers.

Dunbeath Castle, which is still inhabited, exemplifies the French influence prevalent in the reign of Charles I, when military features were being displaced to meet the needs of domestic comfort.

In the interior of the county are the ruins of old Brawl Castle (with walls ten feet thick), the seat of the Earls of Caithness of the Stratherne line, erected in the fourteenth

Keiss Castle and Tower
(the Tower to the Right)

century, and Dirlot Castle, situated like Brawl in the valley of the Thurso and belonging to the same period. Dirlot was a stronghold of the Sutherlands.

The principal municipal and domestic buildings in the county are of recent date. One or two dwelling-houses in Thurso belong to the eighteenth century as well as one or two of the county mansions, but there

E

is nothing in Wick older than the nineteenth century. One house in Shore Street, Thurso, bears the date of 1686. It has a circular turret in front containing the staircase.

Caithness sandstone or flagstone scarcely lends itself to brightness or massiveness, and few of the domestic

Thumster House

dwellings are specially striking. Among the handsomer buildings in Wick are the Commercial Bank (built of freestone), the Court House, erected in 1866, and the North of Scotland and Town and County Bank. Thurso possesses one or two fine public buildings, such as the Miller Institution, the Town Hall, and the Dunbar Hospital. One of the chief architectural features of the county is Thurso Castle, the residence of Sir George

F. S. Sinclair of Ulbster, picturesquely situated at the mouth of the Thurso River. The original building belonging to the seventeenth century was extended and modernised in the nineteenth. Langwell House, near Berriedale, the residence of the Duke of Portland, has its effect embellished by the fine plantations which now surround it.

18. Communications

Caithness now possesses about 300 miles of excellent roads. Until the end of the eighteenth century inland communication was maintained by horse tracks connecting the principal towns and villages. Communication with the south of Scotland was conducted almost entirely by sea. The road from Thurso to Latheron, the first road of any importance in the county, was constructed between 1785 and 1790, under the direction of Sir John Sinclair, by means of statute labour. Thurso Bridge was built in 1800, and in the same year a bridge was built at Wick. Under the provisions of the Highland Roads Act (1803) the " Parliamentary " road from Inverness to Thurso was extended from the Ord to Wick and thence to Thurso. It was completed in 1811. One half of the cost of this important road was provided locally, chiefly by the proprietors and farmers ; even the small tenants had to pay their quota as statute labour had been commuted to a money payment in 1793. Between 1806 and 1860 several new roads were constructed and provision made for the maintenance of existing roads. In 1818 the mail coach, which had already been established between Inverness and Tain, commenced running by Bonarbridge and the Ord to Wick and Thurso, thus opening regular communication by land with the south of Scotland. The County Road Act of 1860 placed all the leading highways under a County Road Trust. Under the older statutes tolls were charged to meet the cost of road repair, but these were abolished by the general Roads and Bridges Act in 1878, when the whole cost of road maintenance was thrown upon the road assessments payable equally by owners and occupiers of heritages. In 1889 Road Boards were constituted under the county councils to administer the maintenance and repair of the county roads. On the development of motor traffic in the early part of this century the tour to John o' Groats became very fashionable, so that for a time the Caithness roads had a heavy motor traffic during the summer, and road repairs formed a heavy burden upon the county. The Development and Road Improvement Funds Act of 1909 provided some relief, and led to an improvement in the roads.

The railway from Helmsdale by Forsinard to Wick (with a branch from Georgemas Junction to Thurso) was completed in 1874. Caithness was thus brought into daily communication with the south of Scotland and the industries of the county, which had made great progress after the construction of the county roads, received a further impetus. The Lybster Light Railway, constructed in 1885, proved of great service to the villages on the east coast, though the benefit would have been greater still had it been possible to carry this line direct from Helmsdale. The detour by rail from Helmsdale to Lybster illustrates the dependence of communications upon the contour of the land. Even now the large section of the county lying to the west of a line from Latheron to Reay is almost destitute of roads on account of natural obstacles. Mail and passenger traffic along the north coast from Thurso to Tongue and Durness, is now conducted by motor power, and the mail coach " Defiance " has disappeared from the road.

19. Roll of Honour

The earliest of the notables of Caithness was the viking Sweyn Asleifson, who flourished in the twelfth century. According to Calder, Sweyn was born in Canisbay and had one of his strongholds on Sinclair Bay. A bold and wily marauder on all the British coasts, he was slain during an attack upon Dublin in 1170. " It has been said," runs the *Saga*, " that he was the greatest man in the western lands, either in old times or at the present day, of those who had not a higher title than he had."

Among the bishops of Caithness several were men of note. To Andrew, the first bishop, a scholar of repute, has been attributed the authorship of the well-known treatise *De situ Albaniæ*. In the seventeenth century Bishop Abernethy was long remembered as the author of *A Christian heavenly treatise containing physicke for the soul.* A contemporary of Bishop Abernethy was the Rev. Timothy Pont, minister of Dunnet, who acquired considerable repute as a map-maker and commentator upon Ptolemy. He died in 1621.

Towards the end of the fifteenth century John Groat, a Fleming, received a royal grant of the lands and ferry of Duncansby. These possessions and the land of Wares and others remained in the Groat family for generations. There were Groats among the heritors of the parish of Canisbay till the eighteenth century. No Caithness name is so familiar throughout the world as " John o' Groat's."

Richard Oswald of Auchencruive, a son of the minister of Watten, is remembered in Caithness through his benefactions for the poor. He acted as plenipotentiary for Great Britain in 1782, when he negotiated with Benjamin Franklin the Treaty of Paris, acknowledging the independence of the United States.

The most notable Caithness man in the eighteenth century was Sir John Sinclair of Ulbster (1754–1835). Born at Thurso Castle, he was educated at the universities of Edinburgh, Glasgow, and Oxford. Next to Adam Smith, under whom he studied economics at Glasgow, he was the most distinguished Scottish economist in the eighteenth century. While taking a leading part in promoting agricultural improvements in Caithness, he also did much for agriculture and fisheries throughout Great Britain. He was one of the founders of the British Fisheries Society and of the Board of Agriculture. The herring-fishing industry at Wick owed its rise mainly to his enterprise and foresight. For nearly thirty years he sat as M.P. for Caithness and Bute. A voluminous writer, he edited the first *Statisti-*

Sir John Sinclair

cal Account of Scotland (1792–98), a work of great national importance, wrote a history of the Public Revenue of Great Britain, and published many pamphlets upon financial questions now forgotten but useful in their day. A biography of Sir John Sinclair was published by his daughter, Miss Catherine Sinclair, herself a voluminous writer of novels and books of travel.

The Rev. John Morrison, D.D. (1750–98), a native of Cairnie, Aberdeenshire, after holding minor educational appointments in Caithness, became in 1780 minister of Canisbay. Possessed of considerable poetical talent, he was a contributor to *Ruddiman's Weekly Magazine* (1771–75). He also prepared historical notes upon Caithness for George Chalmers' *Caledonia*. His title to permanent fame rests on his work as a sacred poet. He was the author of seven scriptural Paraphrases (Nos. 19, 21, 27, 28, 29, 30 and 35) in the collection published by the authority of the Church of Scotland.

One of the worthies of Caithness in the eighteenth century was Sir William Sinclair of Dunbeath (d. 1767), founder of the Baptist community at Keiss. He was the author of a small volume of hymns.

Among well-known sons of Caithness in the nineteenth century was James Bremner, engineer (1784–1856), a native of Keiss, who gained distinction throughout the whole country by his skill in harbour and marine engineering. His fellow citizens erected a statue to his memory on the high ground overlooking the entrance to Wick harbour.

Alexander Henry Rhind of Sibster (1830–61) occupies a high place among Scottish antiquaries. He explored the remains of the neolithic period in his native county. His researches also extended to many districts on the continent of Europe, in North Africa (including Egypt), and in Greece. He bequeathed rich benefactions to Caithness and to Edinburgh University, and founded the Rhind Lectureship in Archæology.

Dr John Rae (1846–1915), a son of Wick, and an eminent writer upon economic subjects, gave lustre to a name long associated with the literature and journalism of the county.   His best known works are *Contemporary Socialism* and the *Life of Adam Smith*.

Mention may be made of two sons of Caithness men who attained distinction in the last century.   Sir Oliver Mowat, a leading Canadian statesman, for twenty-four years premier of Ontario, was the son of John Mowat, a native of Canisbay.   The essayist and historian Henry Duff Traill, though born in London, was connected with the Traills of Rattar.

20. Chief Towns and Villages of Caithness

(The figures in brackets after the names give the population in 1911, and those at the end of each section are references to pages in the text.)

N = Norse.   G = Gaelic.

Berriedale, N. *Bjarg-dalr*, " Stony dale," in the *quoad sacra* parish of the same name, is noted for its fine scenery, being the most richly wooded part of Caithness.   The population of the parish has steadily declined since 1851.   The eldest son of the Earl of Caithness receives the title of Lord Berriedale.   (pp. 16, 23, 25, 57, 64, 67.)

Brough, N. *Bjarg,* " Stony," or G. *Bruach,* a fishing hamlet 3 miles from Dunnet Head, has a small pier.

Castletown (*c.* 900), 5 miles S.E. of Thurso, is the chief village in the parish of Olrig. Its flagstone quarries were at one time considerable. (pp. 3, 14, 44, 60.)

Dunbeath, G. " Birch Hill," on the left bank of the Water of Dunbeath, half a mile from its mouth, and 22 miles south of Wick. There was once a parish of Dunbeath, but it has long been incorporated with Latheron. Fishing, formerly the chief industry, is now in a backward state. (pp. 8, 25, 64.)

Halkirk (*c.* 400), N. *Ha-Kirkja,* " High Church," was the seat of the Bishopric of Caithness until 1222. It is regularly laid out on the banks of the Thurso, about 2 miles from Georgemas Junction. Near it are Brawl Castle and the Combination Poorhouse for the Western Parishes of the County. (pp. 3, 48, 49.)

Keiss (*c.* 340), N. *Quoys,* " Cattle-pen," 8 miles N.W. of Wick, is the chief village in the *quoad sacra* parish of the same name (formed in 1833). On the cliffs near it are the ruins of Keiss Castle. There were 17 boats belonging to Keiss in the year 1911, and 45 resident fishermen. (pp. 22, 25, 42, 60, 64, 72.)

Latheronwheel and **Janetstown** (123), G. *Latharan-Faoilidh,* " Fertile homesteads," about equidistant from Wick and Helmsdale (19 miles), is a fishing village. In 1893 it had over 20 boats, but in 1913 the number was reduced to 10. Near it is the Combination Poorhouse of the Eastern Parishes of Caithness. (pp. 67, 69.)

Lybster (626), N. *Hlith-bolstadr,* " Slope-town," the centre of one of the Scottish fishery districts, is the terminus of the Wick and Lybster Railway. The fishing industry in

Lybster and neighbouring villages has for some years been declining, and the decline applies to the herring fishing as well as to the inshore white fishing. The catch of herrings for the season of 1911, when 70 men and boys engaged in fishing, was only 536 crans. (pp. 3, 25, 42, 45, 68, 69.)

Murkle, N. *Myri-kelda*, " Marsh-spring," a safe harbourage, where at one time the kelp industry was carried on. (p. 22.)

Pulteneytown was founded in 1808 by the British Fishery Society, and named after Sir William Pulteney, its chairman. In 1902 it was united to the Burgh of Wick. The population in 1901 was 5137, while that of Wick was 2774. (p. 45.)

Reay, G. *Ra* or *Ratha*, " circle," at the head of Sandside Bay, abounds with circles and cairns, of the Stone Age. Reay gives his titular name to Lord Reay, the head of the Clan Mackay. The headquarters of the clan were at Reay in the sixteenth and early seventeenth centuries. (pp. 21, 60, 62, 69.)

Scrabster, N. *Skjarr-bolstadr*, " Skerry-town," has an important harbour, which is used both for trading and fishing. In early times Scrabster palace was the northern residence of the bishop of Caithness. (pp. 5, 42.)

Thurso (3335), N. *Thjors-á*, " Bull-river," in early times the chief port in the north of Scotland, for centuries carried on trade with Norway, Denmark, and the Low Countries. When the Continental trade declined, it exported grain and other local products to the south of Scotland and the Hebrides. In 1633 it was erected into a burgh of barony by Sinclair of Ulbster. For nearly two centuries thereafter it was to all intents the county town, although Wick had been erected into a royal burgh in 1589, and had been made head burgh of Caithness in 1641. Beautifully situated in attractive surroundings, Thurso has maintained its position as a prosperous market town, but since 1881 the population (then over 4000) has steadily declined. It possesses several fine public buildings, a public library, museum, and excellent secondary school. An improved harbour would be a boon to the town. (pp. 3, 4, 16, 21, 29, 38, 39, 42, 44, 45, 59, 60, 62, 65, 66, 67, 68, 69, 70.)

Wick (9086), N. *Vík*, " Bay," was in the Norse period of

Wick High School

less importance than Thurso. Though Courts were established here in 1503, and it was made a royal burgh in 1589, and the county town in 1641, yet it remained of small account until the rise of its fisheries towards the end of the eighteenth century. The improvement of the harbour begun in 1810 has continued intermittently since that date. The harbour works have been repeatedly destroyed by heavy seas. A fine new bridge over the Wick River was opened in 1877. (pp. 1, 2, 3, 4, 5, 6, 25, 29, 30, 35, 38, 42, 44, 45, 50, 51, 57, 60, 64, 66, 67, 68, 72, 73.)

SUTHERLAND

SUTHERLAND

1. County and Shire.[1] Origin of Sutherland

In the tenth century the name Sutherland (N. *Sudr-land*, " Southland ") was given by the Norse to the region lying between the Ord and the Oykell, where a number of Norse settlements had been made. This district towards the close of the twelfth century was granted as a barony by William the Lyon to Hugh son of Freskyn of Moray. After Hugh Freskyn had possessed these lands of " Sudrland " for many years he was succeeded by his son William, who about 1235 was created Earl of Sutherland.

For four centuries the name Sutherland was applied only to the south-eastern portion of the modern county. In 1631 the name was extended to the whole country stretching from the Oykell and Kirkaig rivers to Caithness. For a time the Halladale river had formed the north-eastern boundary but in that year the modern boundary, extending from Drumhollistan to the Ord, was finally fixed.

The erection of Sutherland into a county was achieved by Sir Robert Gordon, who was ambitious to raise the prestige of his family above that of such neighbouring barons as the Earl of Caithness and Lord Reay.

[1] See p. 1.

Earlier grants of jurisdiction had been made to the
Earl of Sutherland both by the Crown and by the Earl
of Huntly, Sheriff of Inverness.   In 1503 law courts
were appointed by act of Parliament to be held
at Dornoch.   These courts, however, were established
merely to save litigants the expense of long and danger-

Dornoch

ous journeys to Inverness, for Inverness-shire then included the whole of Ross, Sutherland, and Caithness.
In 1583 the Earl of Huntly made a grant to his relative
the Earl of Sutherland of that part of his jurisdiction
consisting of the sheriffship of Sutherland and Strathnaver.   The Earl of Sutherland thus acquired the jurisdiction of Sheriff over all persons residing upon his own
lands.   In 1601 a royal charter was obtained confirm-

ing to the Earl of Sutherland many old grants and privileges including the regality of Sutherland given in 1347 by David II and the grant of 1583.

In 1631, under a crown writ granted by Charles I, the sheriffdom of Sutherland was severed from Inverness and the lands of the earldom of Sutherland, together with Assynt and the other baronies lying between

Seal of Dornoch

Ross and Caithness, were formed into a shire. Dornoch was appointed to be head burgh of the new shire and the seat of justice in all time coming. The tenacity with which the nobles clung to their feudal privileges is indicated by the fact that in the royal writ of erection special reservation is made to the Earl of Sutherland of the privilege of pit and gallows within his own proper lands as a right apart from the sheriffship. The Scots Parliament confirmed the new erection in 1633.

F

2. General Characteristics

With only ten inhabitants to the square mile Sutherland is the most sparsely populated county in the
British Isles. Except in Lairg and Rogart (which are
inland parishes) the population is confined almost
entirely to the coasts. From Lairg to Cape Wrath, a
distance of fifty miles as the crow flies, the country is
a desolate wilderness. From the top of Ben More, the
highest point in the county, one can behold in every
direction miles upon miles of country destitute of any
sign of human occupation. Unless perchance a few
sheep may be visible in some mountain corrie, nothing
except natural moorland, loch, river, and rock, comes
within the observer's ken. At Duchally, in the upper
reaches of the Cassley river, spots of green appear in
the otherwise limitless expanse of brown heather ;
but these patches of greensward only indicate where
human beings lived about a hundred and twenty years
ago. The characteristic feature of Sutherland is the
vast wilderness which extends throughout its whole
interior—dreary treeless stretches of moorland and
waste.

Another striking feature is its mountain scenery.
Sutherland mountains do not form ranges like those in
some parts of the south Highlands. They are detached
masses, rising separately in positions scattered throughout the limitless wasteland. From any one of these
great mountains most of the others are visible. Striking

Scene near the head of Loch Assynt

(Ardvreck Castle on the peninsula : in the foreground the shell of Calda House)

views are to be obtained in the western parishes of Suilbhein, Canisp, Stack, and Arcuil, though they are perhaps best observed from the deck of a vessel well out in the Minch. Clibreck and Ben Hee also provide a great range and variety of view. The rock scenery on the north and west coasts, whether viewed from sea or from land, forms a striking foreground to the mountain peaks of the interior. The numberless lochs, particularly in the west and north, many of them alive at the proper season with bird life, and full of water lilies and other freshwater plants, also form a notable feature of Sutherland. Only in the south-east of the county are there woods. The moorlands appear all the more dreary owing to the lack of trees and of variety in the plant life. Heather, coarse grasses, cotton grass, and deer's hair moss make up a great part of the inland vegetation. On the high mountains beautiful Alpine flowering plants are found in some of the corries.

3. Size, Shape, and Boundaries

Sutherland is roughly quadrangular in shape, with a projection towards the Ord on its eastern side. It extends some 50 miles from east to west by 40 miles from north to south, and covers an area of 2028 square miles. Excluding inland water the area of the county is 1,297,914 acres, of which two-thirds consist of mountain and heath land used for grazing, and nearly one-third is deer forest. Only 22,661 acres are arable. In 1905 the woodlands covered 20,000 acres, but a great deal of timber has been cut down during the past ten years. The area of inland waters is 47,633 acres. There are 12,812 acres of foreshore and 1553 acres of tidal water. Sutherland is fifth in area among Scottish counties, ranking after the other four Highland counties of Inverness, Argyll, Ross, and Perth. Slightly exceeding Aberdeenshire in size, Sutherland possesses less than four per cent. of the extent of arable land in that county.

The greatest length from east to west within the county is 63 miles, along the line from Rhu Stoer to the Ord. The diagonal line from Cape Wrath to Dornoch Point is also about 63 miles. The greatest length from north to south is 48 miles from Strathy to Dornoch. From Far Aird Head to Oykell Bridge the distance is 45 miles and about the same from Cape Wrath to Cromalt Hills.

On the north and west coasts are several inlets and also bold headlands, such as Strathy Point, Whiten Head, Faraird Head, and Rhu Stoer. The angles of the coast line at Cape Wrath and Dornoch Point give the county a rectangular outline. The south-eastern boundary extends 25 miles along the Moray Firth coast from Dornoch Point to the Ord, and the eastern Caithness border 30 miles from the Ord to Drumhollistan. The Atlantic Ocean forms the northern boundary for over 40 miles, from Drumhollistan to Cape Wrath. The winding coastline following the indentations of Tongue, Eriboll, Durness and the other smaller bays and inlets is fully 80 miles. On the west the Minch coast along a front of 36 miles is rendered irregular by the inlets of Inchard, Laxford, and Loch Cairnbawn and the projection of Rhu Stoer. As on the north coast, these inward and outward windings double the length of the coastline. The southern boundary formed by the Oykell and Kirkaig rivers and the Dornoch Firth, extends 50 miles from Dornoch Point to Kirkaig. Between Loch Borrolan and the Oykell the county of Ross projects into Sutherland as far as Ben More. The burn of " Aultnacealgach " (burn of the deceivers), the boundary north of Loch Borrolan, is associated with a tradition that at the delimitation of the boundary the Ross-shire men secured this tongue of land on the west side of the Oykell by practising a deception upon their Sutherland neighbours.

4. Surface and General Features

The Oykell valley forms a great cleft separating the two northern counties from the southern mainland, so that the earlier geographers named Sutherland and Caithness the " Innismor," or great island. At Invershin this cleft bifurcates into a north-westerly branch running by the Shin valley and Lochs Shin, A'Ghriama, Merkland, More, and Stack to Loch Laxford and a westerly branch trending along the valley of the Oykell and by Lochs Urigil, Veyatie and Fionn to Loch Kirkaig. From north to south Strath Ullie and Strath Halladale form a marked cleft parallel to the eastern border, while at Altnaharra the depression along the Shin and the

Tirry bifurcates into a north-easterly cleavage along Loch Naver and Strathnaver and a north-westerly one by Strathmore and Loch Hope, which strikes the northern ocean at the mouth of Loch Eriboll.

The striking features in the western basin of the county are well seen from Handa. Looking eastwards from the high cliffs of that island, one observes on the left the mountains of the Parph, of which Farmheall is the chief ; Spionnaidh and Cranstackie are seen in the distance, while Foinnebhein and the Reay Forest are in the foreground. Arcuil rises almost due east, with Ben Stack on its right and Ben Leoid beyond Glen Dhu and Glen Coul. To the south-east rises Ben More, with Glasbheinn and Quinag in the foreground, on the right, while behind Quinag is Canisp, and almost due south Suilbheinn and the massive Coigach Mountains. The foreground of this impressive panorama of mountains consists of a wide stretch of bare and hummocky archaean rock, with countless lochs and lochans scattered over its uneven surface.

A good point for observing the eastern mountains is Carnbhren, some miles south of Bonarbridge. From its summit appear, rising from the Moray Firth coast, the whole south-eastern range, including Ben Horn, Ben Dobhran, and the Caithness mountains. Far to the north-east are the two Ben Griams ; Armuinn and Clibreck are outstanding with Ben Hope in the far distance behind, while Ben Hee forms with the bold summits of the Assynt and Coigach Mountains a highly striking outline in the west.

In the vast tracts of uninhabited country wild animals, including game of all kinds, are naturally plentiful. The deer of Durness and the " Dirimore " (great ridge) have been noted for centuries, but to the south of the " Dirimore," Sutherland is better adapted for sheep and cattle than for game.

5. Watersheds, Rivers, and Lakes

The watersheds, being clearly defined, form political as well as natural boundary lines. From the Ord the eastern watershed runs for about 15 miles along the ridge between Sutherland and Caithness, dividing the basin of the Ullie from that of the Langwell and Berriedale. Striking north-west at the Knockfin Heights, it crosses the Kildonan road a mile south of Forsinard, swerves westward for about 6 miles to the source of the Rivigill burn, forming the " Dirimore " between the Mackay country and the Earldom of Sutherland. After running almost due south for 15 miles to the northern slopes of Beinn Armuinn, along the ridge separating the basins of the Brora and the Ullie from that of the Naver, it stretches westward for 20 miles, crossing the Lairg-Durness road at the Crask and striking the western watershed at Meall Garbh some miles beyond Ben Hee.

The western watershed extends from Cape Wrath to the Cromalt Hills. Running south-east and south from the Cape, it strikes the top of Ben Dearg in the Parph, crosses the Durness road at Gualin, passes over

Foinnebheinn, Meallchoin and Sabhal More and, after traversing the road to Scourie midway between Loch Merkland and Loch More, passes southward for 10 miles to the summit of Coinnemheall (Ben More), thence along the ridge of Brebag (where it forms the Ross county boundary) until it crosses the Lochinver road 2 miles east of Aultnacealgach, whence it goes west by south to the Cromalt Hills.

The rivers of this western basin, the Kirkaig, Inver, and Laxford, run short and rapid courses into the Minch. In the Reay basin, north of the " Dirimore," the Halladale (21 m.), Strathy (21 m.), Naver (22 m.), Borgie, Hope, and Dionard (or Grudie) run northerly courses into the Atlantic. The Naver is considered one of the best salmon rivers in the north, while the Halladale and Hope are also noted fishing streams.

In the south-east the Oykell (30 m.), Fleet (14 m.), Brora (27 m.), and Helmsdale or Ullie (27 m.), all flow into the Moray Firth. Rising in the corries of Ben More, the Oykell, after receiving the Cassley and Carron, enlarges into the Dornoch Firth, where it is joined at Skibo by the Evelix. The Fleet flows into Loch Fleet at the Mound, where it is joined by the Torboll burn from Loch Buidhe. The Brora and the Ullie, excellent salmon rivers, drain numerous lochs lying in the region of the watershed, of which Loch a'Chlair, Badanloch, and Loch Ruthair in Strathullie, and Loch Brora in Strath Brora are the largest. In the lower basin of the Oykell are Lochs Buidhe, Laro, Laggan, and Migdale. The largest loch in Sutherland is Loch Shin (21 m.), into which flow the waters of Lochs Merkland and A'Ghriama.

The western basin has hundreds of lochs and lochans well known to sportsmen. The Kirkaig drains Lochs Borrolan, Urigill, Veyatie, and Camloch. The waters of the picturesque Lochs Assynt and Beannoch are carried to the Minch by the Inver, while Lochs More

Loch Migdale

and Stack are drained by the Laxford. Lying in a snug valley between Stack and Arcuil, Loch Stack is reputed the best fishing loch in Sutherland while the Laxford is equally noted as a salmon river.

In the northern basin are Lochs Hope, Loyal, and Naver, each possessing a sentinel mountain by its side. Smaller lochs, though not so numerous as in the western basin, are plentiful all over the Reay country. Fully one hundred lochs are visible from the higher view-points. Loch Na Meide, connected by the Mudale river with Loch Naver, lies near Loch an Dithreibh, which is drained by the Kinloch water into the Kyle of Tongue.

6. Geology [1] and Soil

Western Sutherland has long been considered a classic area by geologists, as there is no other part of the British Isles where the older formations can be so well studied. The effect of rock structure upon scenery is strikingly illustrated in this wild and treeless country, equally attractive on account of its scenic grandeur and its scientific interest.

The archaean gneiss, extending in great hummocky stretches through Durness, Eddrachillis, and Assynt southwards into Coigach is the oldest rock formation in the British Isles and one of the oldest in the world. On this ancient foundation the stratified rocks have been super-imposed. The older groups—Torridon, Cambrian, Silurian and Old Red Sandstone—are disclosed in masses which appear in succession eastwards to the Pentland Firth. Through these formations intrusive masses of igneous rock have penetrated in several places. In the region of Ben Loyal and Ben Stomino this intrusive rock is syenite, resembling granite but containing traces of the dark mineral named hornblende. A mass of grey granite extends between

[1] See pp. 12 to 15.

Loch Migdale and Loch Laggan, near Bonarbridge, where it forms Migdale Rock and Ben Bheallaich. Eastward from Loth and extending into Caithness, as far as Berriedale Water, lies another mass of granite which culminates in Ben Bheallaich near Helmsdale (1940 feet). The most extensive outcrop of granite in the county (also extending into Caithness) reaches from Kinbrace northwards to Sandside Bay and along the valley of the Halladale to Portskerra. The western archæan rock, which has been penetrated by numberless dykes of granite, syenite, and other igneous rocks, rises to a height of nearly 1500 feet in Glasbheinn and Ceannabeinne in Durness, and Ben Stroma in Eddrachillis.

Resting on the gneiss are great masses of Torridon sandstone, which rise into huge mountains of reddish brown conglomerate in nearly horizontal beds of great thickness. Quinag, Canisp, Suilbheinn in Assynt and Fasbheinn in Durness are mainly composed of this ancient rock, while near Cape Wrath sea cliffs of the Torridon rise from the water's edge to a height of nearly 1000 feet. Handa and Rhu Stoer are also Torridonian. Above the Torridon sandstone lie Cambrian formations of white quartzite, limestone, and thin shales, in which occur the fossil named " Olenellus " the discovery of which enabled scientific men to fix the age of the Cambrian. These rocks form a band stretching from the west side of Loch Eriboll to Loch More and thence by the upper Oykell to Ullapool in Ross-shire. To this series belong the limestones of Durness and Assynt, which lie above the " Olenellus " beds and piped quartzites. The largest outcrop of limestone is at Stronechrubie in Assynt, and considerable masses occur at Durness and Eriboll. Owing to great earth disturbance in the Cambrian period quartzite beds have been tilted over the more recent pipe-rock formation, on Cranstackie, Foinnebheinn, Arcuil, Beinn Uidhe, and Ben More.

Silurian metamorphic schists cover the whole of central Sutherland eastwards from the " Great Thrust " to Strath Halladale and Strath Ullie. Along this " Thrust " rocks have been heaved up, overlapped and crushed together by the action of colossal natural forces. Its line can be followed from Whitenhead southward along the

Olenellus

east side of Loch Eriboll till it crosses Loch More and Loch Gorm and, after following the east side of Loch Ailsh, sweeps westward to the Cromalt Hills. The Silurian country is dreary and bare, consisting largely of boggy heather-clad moorland with occasional elevated masses like Ben Hope, Ben Hee, Ben Clibreck, and Beinn Armuinn. On the top of Beinn Armuinn traces occur of the Old Red Sandstone, which was formed above the schists and stretches along the eastern side of the county from Meikle Ferry to Helmsdale—tentacles as it were of the main Old Red formation in Caithness. Similar tentacles on a smaller scale stretch along the north coast to Cnocfreicadain at Tongue.

In a narrow strip between the Old Red Sandstone and the Moray Firth, extending from Golspie to the Ord, are outcrops of a succession of shales, sandstones, coal and limestone belonging to the secondary or newer rocks. At Dunrobin these sandstones and limestones contain a good deal of plant remains, probably carried down by some ancient river to the sea. At various points from Dunrobin to Brora beds of plant refuse have formed into coal, of which the chief bed lies between the upper and lower Oolites at Brora.

Except the boulder clay these beds of Lias and Oolite (Jurassic) are the most recent rocks to be found in the north of Scotland. Resting upon them and stretching over the whole country lie masses of gravel and sand, marking the former presence of ice. Smooth eroded outcrops of gneiss disclose glacial action not merely in the valleys but on the tops of the high mountains.

The metamorphic Silurian schists contain in irregular quantities gold, silver, iron, and other minerals, which occur in various localities. The gold-diggings in the Suisgill and Kildonan burns in 1869 were carried on in the gravel and debris from these schists.

7. Natural History

Embedded in debris in the caves of Alt-nan-uamh near Stronchrubie were found bones of the brown bear, the arctic lynx, the lemming, and other animals now long extinct in Britain. The reindeer, last survival of the fauna of the Stone Age, became extinct about the twelfth century. Sir Robert Gordon, writing of the forests of Sutherland in 1630, says that they are full of "reid deer and roes, woulffs, foxes, wyld catts, brocks, skuyrrells, whittrets, weasels, otters, matrixes, hares, and fumarts." The last wolf in Sutherland was killed in Glen Loth about the end of the seventeenth century. A century later the woodlands of the county being very limited, the squirrel became extinct, reappearing, however, about 1860 in the new Sutherland plantations and now so abundant that it has to be steadily hunted down. Between 1819 and 1826, when proprietors and sheep farmers offered premiums for the destruction of "vermin," 1143 wild cats, martens, and polecats were killed in Sutherland and on the Caithness border, with the result that the polecat, marten, wild cat, and stoat are now rare. The fox, though persecuted, is still fairly common. The badger is said to occur at Ben Bhraggie but verges upon extinction. The otter occurs in various eastern streams, such as the Brora and the Spinningdale burn. Sir Robert Gordon maintained that rats could not live in Sutherland, but the brown rat is now very abundant. The water-vole is common, and the field-vole, particularly in limestone districts. The brown hare and mountain hare, though not so common as formerly, are found all over the country. The rabbit is plentiful in sandy districts and wherever there is suitable cover.

Sir Robert Gordon

Roe deer abound in the woodlands of the south-east. The common seal frequents the shoals and sandbanks of Little Ferry and the Dornoch Firth. Both the grey seal and the common seal occur on the west and north coasts.

The bird life of the county possesses great variety and interest. The osprey, kite, and merlin, are extinct, and the sea eagle now rarely breeds, though formerly fairly common on the west coast. The capercailzie, common in the seventeenth century, has long been extinct, and an attempt about 1870 to reintroduce it at Skibo was unsuccessful. The great northern, red-throated, and black-throated divers occur in the numerous lochans of the western parishes. In the boggy wilds of central Sutherland nest black-headed gulls, many kinds of wild duck, the coot and the dunlin, while the plover and wild goose lay their eggs in the dry hummocks and knolls near such waters as Badanloch. The marshes above the Mound provide cover to numerous water-fowl such as herons, oyster-catchers, and waders of every kind, who feed at low tide on the flats of Loch Fleet. Assynt contains about three hundred lochs, many of them alive in the season with such water-fowl as greenshanks, dunlin, and heron, while the call of the curlew is familiar on the moors. Ptarmigan, which have disappeared from the Griams and other eastern hills, are fairly numerous on Clibreck and the higher Assynt mountains.

Among summer visitors who come in numbers are the wheat-ear, whinchat, red start, white throat, willow

G

wren, wood wren, sedge warbler, grey wagtail, ringed plover, fly-catcher, night jar, arctic tern, and lesser tern. The redwing, fieldfare, shrike, and snow bunting are autumn visitors. The latter has been known to nest in some high corries of the western mountains. Among rarer resident birds of the county are the raven, magpie, cross-bill, siskin, bullfinch, goldfinch, buzzard, and grebe. Owls are common in the Skibo and Dunrobin woods. The grey lag goose and whooper swan are winter visitants. The eider duck breeds at Eilanhoan (Loch Eriboll) and the goosander in Loch Shin.

The adder and the lizard are found all over the county, the slow worm and newt in the western parishes.

Salmon, charr, sea trout, and river trout occur in most of the lochs and streams. Among sea fishes large halibut are found off Cape Wrath, while cod, ling, plaice, turbot, and sole are everywhere abundant. Sand eels (lesser launce) are common in the tidal shoals and sandbanks. Crustaceans such as lobster, crab, and mussel abound along the rocky coasts in the north and west, while clam and cockle are found at low tide on the sandy flats. The streams in the limestone country abound with the lower forms of crustacea. Whelks, limpets, mussels and other molluscs abound on the rocks lying between low and high water-mark.

Alpine vegetation on the higher mountains represents the flora of the period which followed the Ice Age. As the climate softened, the flora of the temperate regions gradually extended over the river valleys reaching to the lower hills while the arctic flora receded to the corries in the mountains. There are still found such arctic plants as *Saxifraga oppositifolia, S. hypnoides, Luzula spicata, Cochlearia, Eriophorum angustifolium* and *Lycopodium alpinium.* Among rarer plants in the higher mountains is *Luzula arcuata,* found on Ben More and Foinnebhein. In some spots arctic vegetation still occurs on the lower grounds and after a succession of cold seasons the upland flora creeps lower down the mountain sides. *Eriophorum pubescens* occurs near Oykell Bridge and *Draba incana* as well as *Saxifraga oppositifolia* in Farr and on the borders of Loch Eriboll. In many of the lochans of the western parishes are found the bulrush, the white water lily (*Nymphaea alba*), and the prickly twig rush (*Cladium mariscum*).

The limestones of Assynt and Durness yield a rich and characteristic vegetation including *Scolopendrium vulgare* and the filmy fern (*Hymenophyllum wilsonii*), which occurs near Inchnadamph, while *Dryas octopetala* occurs at Stronchrubie and Smoo.

The vegetation of the inland moors and uplands largely comprises heather (*Calluna vulgaris*), deer's hair moss (*Pleocharis caespitosa*) and cotton grass. At times patches of other plants occur to break the dreary uniformity of the moors. Near Loch Craggy in Assynt, for example, Anderson observed quantities of *Carex uniflora*; *Ribes petraeum* occurs at Rosehall, where also *Pinguicula lusitanica* and *Drosera anglica* abound, while in the low marshes of Oykellside occur *Malaxis paludosa, Pilularia, Globulifera* and *Nymphaea alba.* At Faraird Head *Scilla verna* and *Primula scotica* occur. *Hieracium denticulatum* flourishes near Oykell Bridge.

Most of the woodlands of Sutherland grow in boulder clay overlying Old Red Sandstone. Scots pine, larch, and spruce prevail but there is also a considerable wood of oak and beech between Spinningdale and Creich and some fine deciduous trees at Skibo and Ospisdale. The richness of the secondary soils of the Dunrobin district accounts for the luxuriance of the vegetation. Though Dunrobin gardens lie close to the sea, yet even in Sir Robert Gordon's time there were " all kynds of froots, hearbs and floors, used in this Kingdom, and abundance of good saphron, tabacco and rosemarie."

8. The Coast Line: Coastal Gains and Losses

On the west the coast scenery possesses striking grandeur. From Kinlochbervie to Cape Wrath the bold and lofty cliffs rising from the water's edge form the haunts and breeding places of great numbers of gulls, cormorants, guillemots, razor-bills, and other sea birds. The numerous islands between Loch Inchard and Stoer, of which the largest are Oldany and Handa, give scenic variety and serve to shelter the coast. Lochinver, Badcall Bay, Loch Laxford, and Loch Inchard provide harbourage for large steamers while smaller craft use Loch Kirkaig and the bays of Stoer, Culkein, Clashnessie, Nedd, Scourie, and Kinlochbervie. The largest inlet is Loch A' Chairn Bhain separating

Cape Wrath

(Showing Archaean Gneiss)

Assynt from Eddrachillis.   This loch narrows at Kylesku
to less than 400 yards, then expands into two branches,
Lochs Glendhu and Glencoul, each penetrating some
10 miles into the land.

The north coast, besides the larger Kyles, possesses
smaller inlets still frequented by fishing craft, such as
Sango Bay and Rispond Bay, between the Kyles of
Durness and Eriboll, and the bays of Torrisdale, Farr,
Swordly, Kirtomy, Armadale, Strathy, and Bighouse,
between the Kyle of Tongue and Drumhollistan.   Loch
Eriboll, a deep water loch along its whole extent, forms
a useful harbour of refuge for shipping.   Though islands
are less numerous on the north than on the west, there
are a few off the Kyles of Eriboll and Tongue.   On
Eilan Hoan at the mouth of Loch Eriboll and on Corrie-
Eilan towards the south end of that inlet are situated
old burial grounds.   Like those on Oldany and Handa,
these date back to a time when the wolves infesting
Reay Forest and the Assynt Mountains prowled in
winter about the churchyards.   Naomh Eilan (the Isle
of Saints) lies off Torrisdale Bay.   One of the natural
wonders of the north coast is the cave of Smoo at
Durness.

The only indentation of the south-east coast line is
the shallow estuary of Loch Fleet.   No bold rocky
headlands, or stacks, occur on the sandy south-east
coast.   Dornoch Point and the headland of Ardnacalc
are simply projecting banks of sand, while the banks
known as the " gizzen brigs " block the entrance to
Dornoch Firth.

The rugged western and northern coasts have been little affected since the glacial period by upheaval or subsidence. Oldany, Handa, and the other western islands are the summits of gigantic elevations rising from the archaean rock bed which was submerged in some remote geological epoch. The south-east coast is mainly composed of Old Red Sandstone and later rocks worn to sand by denudation. Raised beaches run along this coast from Helmsdale to Bonarbridge, indicating upheaval of the land in or since the glacial period. The extensive tidal banks in the outer Firth of Dornoch and in Spinningdale Bay were formed partly by upheaval followed by extensive denudation and partly by accumulation of silt. Sandbanks in the Kyles of Tongue and Durness have been formed in the same way. Oolite and other newer rocks along the seaboard of Clyne have also been wasted into shelving beaches of sand. The older rock formations provide numerous good natural harbours on the west and north coasts, while on the sandy Moray Firth coast harbour accommodation has been a standing difficulty. Embo, the chief fishing village in the county in point of population, has but paltry fishing returns mainly owing to the want of a decent harbour.

When the Mound was constructed in 1811–12, a considerable stretch of haugh land was recovered from the sea. Soon afterwards the Lonemore of Dornoch was drained and the lochs and bogs of Crakaig in Loth were reclaimed so as to form one of the best farms in the county.

The lights at Cape Wrath and Stoerhead serve for the navigation of the Minch. Cape Wrath light is an alternating white and red light visible for 27 miles. The light on Stoerhead, which is a white light visible for sixty seconds and eclipsed for thirty seconds alternately, has a range of 20 miles. The main light on the east coast is at Tarbetness in Ross, but there are minor lights at the piers of Embo and Golspie (lighted when boats are at sea); at the north side of Little Ferry (visible for 3 miles); and at the north pier of Helmsdale.

9. Climate [1]

One of the most striking features in the climate of Sutherland is the contrast between the mild humidity of the western basin, with its balmy Atlantic breezes and the cold northerly and north-easterly winds experienced in the northern and eastern districts, particularly during the spring months. This humidity arises from the fact that the vapour-laden Atlantic winds become cooled on striking the western hills and the vapour condenses into rain. Westerly winds prevail in winter and spring, producing an average annual rainfall in the north-west districts of 60 inches, while in the northern district it is 36 inches and in the southeast only 31 inches. In the west rain falls on an average upon two hundred days in the year. On the other hand Dornoch and the south-east of the county is

[1] See pp. 29-32.

reckoned to be within the dry belt of eastern Scotland. Owing to the prolonged spring winds the mean annual temperature in this district is only about 45° F.

The summer in Sutherland is short, and early autumn frosts often prove harmful to crops, especially in those narrow glens where the high mountains exclude sunshine for a great part of the day. Mr James Loch states that early autumn frosts caused havoc to potato and corn crops in the old days before rotation. On the high lands of the interior winter is frequently severe. Snow lies long not only on the hills but even on the lower ridges, so that the Crask has ever been a terror to travellers in the winter season.

10. People : Race, Language, and Population

The story of the early races in Sutherland—from primeval days to the coming of the Scots—is practically the same as in Caithness.[1]

The Scots, who had settled in Argyll and the Isles, gradually spread northwards and eastwards into the country beyond the Oykell. Though Gaelic was introduced into Argyll within the Roman period, probably it did not extend into Sutherland until the seventh century, when the Columban missionaries settled among the northern Picts. By the Columban missionaries Gaelic was employed in religious services and a

[1] See pp. 33, 34.

knowledge of the language was communicated to the native Picts, whose ancestral tongue gradually became extinct. Gaelic supplanted Pictish much in the same way as English is now supplanting Gaelic. At the arrival of the Norse in the ninth century Gaelic was the language of the people. A large proportion of the place names then in use were Gaelic, though a few

Dunrobin Castle in 1812

Pictish names of places (such as Pitgrudy and Pitfour) still survive in the county. In the thirteenth century Bishop Gilbert translated the Psalms into Gaelic for the use of the people. In 1544 at the Baillie Court at Dunrobin letters of charge were read to the tenants of the earldom enjoining them to pay terce to the dowager-countess. The letters were first read in English and then explained to the people in Gaelic by interpreters.

A century ago Gaelic was still the language in general use in the county though English was the official language. The first inflow of a goodly number of English-speaking people took place after 1809, and by 1871 more than one-fifth of the people spoke English only. The introduction of board schools and the construction of the Sutherland Railway after 1872 caused a rapid spread of English. In 1911 out of a total population of 20,179 there were 11,651 persons acquainted with Gaelic and English, and 188 acquainted with Gaelic only.

Fully one-fifth of the present population of the county is extraneous. In 1911 there were 4201 persons enumerated whose birthplace was outwith the county, compared with 3957 in the year 1901. The extraneous population is increasing while the indigenous is steadily decreasing, at an even greater rate than the increase of the other class.

There have been within the historic period several migrations into Sutherland from other parts of Scotland. The Norse, who settled in the tenth and eleventh centuries, occupied the best arable land on the east coast and along the Oykell, while they also possessed a few settlements on the north and west coasts. Centuries later the Gunns, a tribe of Norse extraction, migrated from Caithness into Kildonan. In the reign of David I the Mackays, after the rebellion of Malcolm Macheth, migrated from Moray and settled in Strathnaver and Reay. A later migration from Moray took place at the end of the twelfth century, when the Murrays were brought over by Hugo, son of Freskyn de Moravia. The Murrays, the native Sutherlands and the Mackays remained the leading families in the county until the early part of the sixteenth century, when the Gordons came over from Strathbogie in the train of Adam Gordon, son of Lord Huntly, who on his marriage in 1509 to the Countess of Sutherland assumed the title of Earl of Sutherland. Between 1815 and 1830 came an inflow of shepherds and border farmers from the south of Scotland as a result of the introduction of pastoral farming into the county. A century ago the population of Sutherland consisted almost entirely of those whose ancestors had occupied lands in it for many generations, while to-day the indigenous inhabitants form little more than half of the population.

Sutherland is the most thinly peopled county in the British Isles, having only ten persons to the square mile, while Inverness-shire, the next most sparsely peopled county, contains twenty-one to the square mile. While Sutherland is fifth in size among Scottish counties in population it is twenty-eighth.

11. Agriculture

Sutherland is, and always has been, a pastoral rather than an agricultural county. Even in the Bronze Period, many centuries before the Christian Era, Lairg, Rogart, Kildonan, and Farr were occupied by a considerable population, but it is scarcely possible that these primitive people could have had much land in cultivation. Composed largely of archaic rocks, the soil of Sutherland contains but little of the constituents that contribute to fertility. On the Old Red Sandstone and secondary rocks of the east coast there is a stretch of good arable land, and some of the haughs in the valleys have been in cultivation from a remote period. Prior to 1810 the common stock of the county consisted of highland garrons, black cattle, and " caory " sheep (now almost extinct except in one or two of the Orkney Islands) while the goat, now so rare, was common. Improved breeds of sheep, cattle, and horses, were gradually introduced during the earlier half of last century and before 1850 a good average stock of the standard breeds and crosses could be seen on all the larger farms. The cattle and sheep of the smaller tenants remained somewhat poor in quality, but the past twenty years have witnessed a marked improvement in all kinds of live stock. Swine are on the increase, although not kept in such numbers as might be expected in a county of smallholders.

In 1912 there were 2612 horses and 11,449 cattle in the county, a slight decrease from the numbers of the previous year. Foals and calves, however, showed an increase. The number of sheep was 215,124, being 2000 more than in 1911. Notwithstanding the enormous extent of its rough pastures Sutherland ranks twelfth among the sheep-rearing counties. Its low relative position is due to the fact that an area of 427,548 acres has been converted into deer forest. There were in 1912 only 809 pigs in Sutherland compared with 15,266 in Wigtown. Small as this number is, it was 88 more than in 1911. Less than the fortieth part of the county is under cultivation. The extent of arable land in 1912 was 22,984 acres, being 2 per cent. of the total area of the county and 1000 acres less than the arable acreage of 1911. Though so small a proportion of the land is arable, there is yet a shrinking of the area in cultivation, part of it being turned into permanent pasture.

Before the fall in the price of wheat, that cereal was grown upon several farms in Sutherland, but in 1912 only a few acres were under wheat. The chief grain crop is oats, which in 1912 was grown upon 8030 acres with a yield of 31,493 quarters, being nearly four quarters to the acre. This is rather below the average yield. The barley crop comes next in importance. In 1912 763 acres were under barley and bere, yielding 3113 quarters or slightly over four quarters to the acre. One-fifth of this crop consists of bere. Rye, at one time extensively grown on lighter soils to provide straw for thatching, is no longer much grown now that roofs of slate and iron have become general, and in 1912 occupied only 65 acres. Green crops consist almost entirely of turnips and potatoes. In an average year there would be about 3000 acres of turnips and 1400 acres of potatoes in the county, amounting together to one-fifth of the arable area. The average area under cereals may be put at 8800 acres, and the area under rotation grasses would be about the same.

The yield of potatoes is about 5 tons, and of turnips

15 tons to the acre. Increased use of fertilizers by the smaller tenants has caused a steady rise in the average yield of green crops. In the cultivation of the land and in the breeding of stock there has recently been a striking advance in many parts of this county. Equally striking has been the improvement in dwellings, farm offices, and farm implements.

Sutherland possesses 2550 holdings exceeding one acre. Of these only 100 exceed 50 acres, while there are 1500 below 5 acres. The holdings between 5 and 15 acres number 775, while there are 144 between 15 and 30 acres. Only 28 smallholdings exceed 30 acres. The fact that 96 per cent. of the holdings in the county are smallholdings and that the vast majority of these are less than 5 acres in extent has a profound bearing upon the social condition of the people. Only 52 smallholdings are occupied by the owners, and there are not 60 persons in the county who own more than one acre of land.

In 1912 there were 9212 acres of woodland in the parishes of Dornoch, Golspie, and Rogart, being 8 per cent. of their total area. At the same time there were 6000 acres of woodland in the parish of Creich. At Dunrobin and Skibo are some fine specimens of oak, beech, and elm, as well as many handsome spruces and other conifers. A large part of the county devoted to deer forests is reckoned to be suitable for afforestation.

12. Industries, Manufactures, Mines, and Minerals

At the end of the eighteenth century George Dempster of Skibo established a cotton mill at Spinningdale, which was burnt down in 1808, and since that time no factories of any importance have been established in the county.

According to the census of 1911 upwards of 3000 persons (male and female) were engaged in agricultural, pastoral, and forestry employments in Sutherland, while there were about 1000 fishermen, 200 curers and fish workers, and 30 coopers. The other industries in the county are mostly accessory to those of agriculture and fishing. The various building trades provide work for about 400 men, while 200 persons are employed in the clothing industries, and other 200 in the industries concerned in the provision of food and drink. In 1911 there were 280 general labourers and 790 female domestic servants.

Brora, the only manufacturing and mining centre, provides employment for 30 men in its coal mine, about 20 in its brick works, and 6 in its distillery. One hundred and fifty men are employed on railways and over 100 on roads. Forty men and 16 women were returned in 1911 as workers in wool and hosiery, chiefly at Brora and Rogart, but this does not include workers in homespuns, who are numerous in Rogart, Lairg, and the north-western parishes. Game preserving afforded employment in 1911 to 182 persons, but prosecutions for poaching are rare, sporting rights of moor and river alike being generally respected by the people. At the last census 1620 males and 6249 females, being one-fifth of the total number of males and three-fourths of the total number

Spinningdale

of females above ten years of age, were returned as of no specified occupation.

The mineral resources of Sutherland, though not lacking in variety, are commercially of little account. Accumulations of slag, the remains of ancient iron workings in the period of the brochs, are to be found along river banks and burn sides in various districts. Such slag heaps occur in Achinduich, Shinness, Lairg, Skelpick —where the layer of slag is intermixed with burnt wood and charcoal—and Achrimsdale in Kildonan

H

Lady Jean Gordon

Sir Robert Gordon affirms that in his time abundance of good iron was made in Sutherland.

For centuries no attempt was made to develop the mineral resources of the county. At length, Jean, Countess of Sutherland, in 1598 opened coal pits and salt pans at Brora. According to her son, Sir Robert Gordon, the Brora salt pans not only served local needs but also provided an export trade to England and elsewhere. After a time, however, coal mining and salt making were discontinued. They were revived for some years about 1820, and the third Duke of Sutherland, about 1870, again opened the coal mine at Brora. Since his time a steady, if not large, trade has been maintained.

In recent years marble was quarried at Assynt and exported from Lochinver, but the venture proved unsuccessful. Extensive granite quarries opened at Dalmore, Rogart, in 1885 with the object of procuring blocks for polishing also proved unsuccessful. An excellent building stone was obtained but the promoters found that the granite did not take on a flawless polish. Granite building stone is quarried at Creich and Lairg, and sandstone at Dornoch and Brora. The hard whinstone found all through the county is used to a considerable extent locally for making up the road crusts. Were the limestones of Durness and Assynt more accessible they might be more extensively utilized by agriculturists. The limekilns in use some thirty years ago at Lairg have been discontinued.

13. Fisheries and Fishing Stations

The Sutherland fisheries fall within three of the twenty-seven districts recognized by the Fishery Board. The east coast fisheries are in the Helmsdale district, those of the north coast belong to Wick, while the west coast stations belong to Lochbroom.

In 1914 there were 51 registered fishing boats in Helmsdale village, 11 belonging to the first class. Embo possessed 25 boats, 6 being in the first class. Golspie possessed 21 boats and Brora 13. Embo had 130 fishermen, Helmsdale 106. The total number of fishermen in Helmsdale district in 1914 was 360. On the east coast (where the haddock and plaice fishings are the most valuable) the yield in 1914 was £6000 from the white fishing and £800 from shell fish.

To the Wick district belong the north coast villages of Portskerra, Strathy, Kirtomy, Skerry, Isle Roan, Talmine, and Eriboll. The boats of the north coast fishermen are nearly all of the small class (under 18 feet keel). The total number of resident fishermen in these villages in 1914 was 210, of whom 40 were in Portskerra, the only village which showed a slight increase on the figures of the previous year. The value of the cod and saith fishing in 1914 was £800, while lobsters and crab yielded £1500.

The west coast fishings are chiefly at Oldshoremore, Badcall, Scourie, Culkein, Stoer, and Lochinver. The fish obtained are chiefly cod, haddocks, lobsters and some herrings. There are 330 fishermen engaged in the west coast fishings, of whom 90 are in Stoer and 30 in Oldshoremore. The annual value of the white

Fishwife, Embo

fishings exceeds £2000, while the yield of the lobster and crab fishings is rather less than £2000.

Owing to lack of improved gear and appliances the Sutherland fishings are on the whole in a backward condition. Fishermen depend for their livelihood not so much on the yield of the local fisheries as upon their earnings at the Shetland, East Coast, and English herring fishing.

The salmon fisheries of Sutherland have been steadily growing in value. The assessed rentals of the Kyle of Sutherland fishings have almost doubled within the last thirty years and now exceed £4500. The rentals of the salmon fishings of the Halladale, Naver, Borgie, Kinloch, Hope, and Dionard, amount to about £2000, but as the angling is often let with the shootings the detailed valuations cannot be definitely stated. The rod fishings of the Borgie are worth about £100 per annum. The number of fish taken on the north coast fishings in 1914 amounted to 2000 salmon, 3000 grilse, and 500 sea trout. There are valuable salmon fishings on the west coast, particularly on the Laxford, the Inver, and the Kirkaig rivers.

14. Shipping and Trade

Prior to the extension of the Highland Railway from Ardgay to Thurso the needed supplies for the county were obtained by sea. Sloops and schooners traded from Leith and Newcastle to Helmsdale, Brora, Little Ferry, and Bonarbridge on the east coast, and on the west side from Glasgow to Lochinver, Badcall, Loch Inchard, and Loch Eriboll. For a good many years a small steamer plied between Burghead and the Little Ferry. With the advent of railways, however, this coasting trade declined. At Bonarbridge and Little Ferry small vessels still unload from time to time cargoes of coal and building materials and take away pit props and timber logs. Brora and Helmsdale also import timber and slate, and export fish (principally herrings). Dunrobin pier is private, while Golspie pier is used mainly by fishermen.

The trade on the north and west coasts is of small account. Some tourist traffic is brought by the west coast steamers to Lochinver, Badcall, and Loch Eriboll. For some years in the middle of last century the fish-curing industry at Lochinver, and the marble quarries at Drumbeg provided shipping freights at Lochinver, but both these industries are now extinct.

Loch Eriboll is in winter a valuable harbour of refuge for steam trawlers and other fishing craft. In the days of sailing ships its waters were sometimes crowded with vessels during severe winter storms. While shipping at Lochinver and Helmsdale is no longer what it was, yet there is a good prospect of increasing trade at the numerous fishing ports and piers round the Sutherland coasts when the fishermen learn to apply to their calling those improved technical methods which have revolutionized the fishing industry in other parts of Scotland.

15. History of the County

Prior to the Norse period the country north of the Dornoch Firth was occupied by a Gaelic-speaking people, among whom Gaelic law and customs prevailed. No written records of this early period have survived the ravages of the Northmen. After their arrival in 872 the Gaelic-speaking people gradually receded from the south-eastern coast and from the river valleys into the interior. While the place names show that the low-lying land near the sea and many of the places in the Oykell valley fell into the hands of the Norse, the

Earl's Cross

interior of the county has remained in possession of Gaelic-speaking people until the present day. The Earl's Cross at Embo is said to mark the site of a battle against invading Norsemen, in which Sir Richard Murray was killed, in 1259.

The organization of the Roman Church was introduced about 1150 when Bishop Andrew was appointed

to the northern see. The bishop and his immediate
successors established themselves in the northern and
wealthier section of the new diocese, where Halkirk
was made the ecclesiastical centre.   When Jarl Harold

Bishop's Seal

was forced to make submission to William the Lyon
in 1202, the Scottish king strengthened the central
power in the north by granting the south-eastern
portion of the diocese to Hugh, son of Freskyn of
Moray, a powerful magnate on the southern shores of
the Moray Firth.   From that time the Murrays steadily

extended their power and Norse influence in that
quarter gradually declined. Meanwhile the vast
northern and western portions of what is now known
as Sutherland were ruled by Celtic maormers. After
the martyrdom of Bishop Adam in 1220 the power of
the Murrays was increased by the appointment of arch-

Bishop's Palace, Dornoch

deacon Gilbert of Moray to the vacant see. He transferred the bishop's seat from the Norse centre at Halkirk
to Dornoch and erected his splendid new foundation
as far as possible from the fierce Norsemen.

In the course of the thirteenth and fourteenth
centuries feudal law and custom gradually replaced
the earlier Celtic jurisdictions not only in the barony
of Sutherland but also in Strathnaver and Assynt.

The Earl of Sutherland, Mackay of Farr, and MacLeod of Assynt were the chief barons of the country. Much of the history of Sutherland from the thirteenth to the end of the seventeenth century is made up of the rivalries and conflicts of these feudal lords. The royal power was of so little account that the whole north of Scotland beyond the Spey formed but a single sheriffdom. Not until 1503 was the King's Court established at Dornoch,

| Seal of William, | Seal of John, |
| Fifth Earl of Sutherland | Eighth Earl of Sutherland |

though no doubt feudal law of a sort was administered by the bailies of the barons. In 1508 the Earl of Huntly became heritable sheriff of Inverness. The following year Adam Gordon of Aboyne married Elizabeth Sutherland, heiress of Earl John. The rise of the Gordons in Sutherland was marked by a further increase in the number of feudal conflicts then so prevalent.

One of the fiercest of these took place at Tutim in

Strathoykell in 1408. Angus Mackay of Farr had married a sister of MacLeod of Lewis. After Angus's death his widow appears to have been maltreated by his son Hugh, and MacLeod of Lewis invaded Strathnaver with a large following of clansmen to protect his sister. The MacLeods of Lewis, having ravaged the country of the Mackays, were pursued and overtaken at Tutim. A fierce battle ensued. According to Sir Robert Gordon, only one of the Macleods, grievously wounded, was able to return to Lewis to relate that Malcolm MacLeod, chief of the clan, and all the rest of his men had been slain at Tutim.

Another fierce conflict occurred in 1427 at Drumnacoub, near Tongue. Thomas Mackay of Creich, son of Neil Mackay, who had fought so valiantly at Tutim, killed Mowat of Freswick within the sanctuary of St Duffus at Tain. In consequence of this crime Thomas Mackay was outlawed. Angus Murray of Culbin, who succeeded in arresting the outlaw, was rewarded by grants of Mackay's forfeited lands in Strathnaver. Murray, who invaded Strathnaver to secure possession of these lands for behoof of his sons-in-law, Neil and Morgan Mackay, met his foes at Drumnacoub. The fight, says Sir Robert Gordon, was so bitter that few survived on either side. The leader of the Mackays (John Aberich) "seemed to have the victory because he escaped with his life," so that Murray's invasion proved abortive.

In 1570 a feud arose between the Murrays of Dornoch and the Sutherlands of Skelbo, whom the Murrays defeated in a skirmish at Torranroy, near Dornoch. To avenge this defeat the Sutherlands called to their aid the Sinclairs under the Master of Caithness, and the Mackays under Mackay of Strathnaver. Sinclairs, Mackays, and Sutherlands attacked the Murrays in Dornoch, burnt the cathedral and the bishop's palace, and ravaged the neighbouring country.

For many years at this period the Earls of Sutherland and Caithness were in fierce conflict about questions of jurisdiction.

In his march through Sutherland in 1651 to the fatal field of Carbisdale, Montrose received scant local support. After his defeat he wandered for days in the wilds of upper Strathoykell until compelled to surrender. When he was taken to Ardvreck Castle, MacLeod of Assynt was from home at the time, but his wife, a strong Covenanter, at once reported his capture to the authorities and Montrose was transferred to Skibo Castle, whence he was taken to Edinburgh. Though Macleod had no personal responsibility for the surrender of Montrose, yet the royalists never forgave him. After the Restoration he was compelled, owing to persecution, to sell his estates. The members of his family took refuge in the Netherlands, where their descendants are still to be found. At the same period several of the Sutherland lairds, including Robert Gray of Creich, Gray of Skibo, and Murray of Pulrossie, were subjected to heavy fines because they had taken sides against the royalists.

In the eighteenth century Sutherland remained

loyal. In the rebellions of '15 and '45 the Earl of Sutherland took up arms against the Stuarts. Though the heritable jurisdiction was lost in 1747, the power of the family was greatly increased by the marriage in 1785 of the Countess Elizabeth to Earl Gower, afterwards Marquis of Stafford and first Duke of Sutherland. In the forty years after 1785 the Sutherland family made large purchases of land and property in the county until, by the purchase of Lord Reay's estates in 1829, the family had acquired practically the whole county except Skibo, Rosehall and Achany.

In 1811 the Countess Elizabeth, then Marchioness· of Stafford, introduced extensive schemes for the improvement of the agriculture of the county, the making of roads, the plantation of woods, the creation of sheep farms and the establishment of a number of fishing villages. These schemes were unfortunately marred by acts of harshness towards small agricultural tenants particularly in Strathnaver and Kildonan.

After 1840 the wilds of Sutherland began to attract the attention of sportsmen and tourists. The extension in 1870 of the Highland Railway system to the county increased its value for sporting purposes. Since that date a very large area has been converted into deer forest.

Meanwhile the Crofters Act (1886) by providing security of tenure to the smaller tenants led to a great improvement in their dwellings and in their methods of cultivation. The benefits of this statute were further extended in 1911.

16. Antiquities

The oldest remains of man and his works in Sutherland
consist of stone cairns extending some sixty to seventy
yards in length, and having projections or horns at each
end. Three examples of these occur in Strathnaver,
two at Ach-coyle-nam-borgie, and one at Skelpick.
In Kildonan, near Helmsdale, there are seven long
cairns and another alongside the railway at Lothbeg,
but in these the horns are wanting. There are also
two unhorned long cairns at the Hill of Kinbrace and
one on the west slope of Cambusmore Hill (Creag an
Amalaidh) near the Mound. These cairns are sepulchral
monuments of the inhabitants of the neolithic period.
They usually contain a chamber partitioned into two
and, in some cases, into three divisions, containing
human remains and such relics as stone implements,
fragments of pottery, the bones of domestic animals,
of reindeer and of other wild animals.

Horned circular cairns occur at three places near
Spinningdale. One of them in the wood opposite the
small farm of Cnocdubh was excavated some years ago
by Mr Curle, who discovered burnt human remains,
one or two relics of pottery, and a short tanged flint
scraper, now preserved in the National Museum at
Edinburgh. The chamber in the cairn at Cnocdubh
is built of seven large slabs placed on end. It is floored
with flags and has a covering flagstone, over which
(before excavation) was piled a cairn of stones 8 feet

high. The other horned round cairns near Spinningdale are situated one in Coiloag Wood—near the game-
keeper's cottage—and the other at the west end of
Ledmore Wood. There is also a horned round cairn
at Aberscross near the Mound station, another on
Lairg Moor, and there are examples in the parishes of
Kildonan and Farr. One or two round cairns without
horns, but containing burial chambers, are to be found
in nearly every parish in the county. These also
would appear to belong to the Stone Age, but are
probably more recent than the horned cairns and
may mark a transition to the Bronze Period.

The cairns testify to the skill as builders of the neolithic people and point to some kind of social organization among them. Near the large cairns there usually
occur a number of smaller cairns or mounds—probably
the burial places of the members of the tribe. The
implements of stone were manufactured with taste as
well as skill. Except the cairns and stone implements
no other monuments survive of the neolithic men, yet
it would seem that this Mediterranean race was far
from being extirpated by the successive hordes of invaders who after them possessed the land. They
migrated northwards along the east side of the country,
and few traces of them survive in the mountainous and
western parts of Sutherland. It is remarkable that in
the remotest past the inhabitants of the county settled
in the districts where to-day population is chiefly to be
found. The late Dr Joass, Golspie, collected a considerable number of relics of the Bronze Age (1500 B.C.–

1000 B.C.) in various parts of Sutherland, and these are now preserved in the Dunrobin Museum and in the National Museum at Edinburgh. In the Bronze Period a race of people from central Europe spread over the British Isles introducing a more advanced civilization than that of the neolithic people. Their burial customs were simpler. The remains of their dead are found in stone cists, sometimes in small cairns, but frequently in the earth a few feet below the surface. Some years ago on the farm of Little Creich the covering slab of a stone cist was turned up by the plough. One or two cairns of the Bronze Period, with cists, occur in nearly every parish on the east side of the county. One cairn has been excavated and the cist exposed at Tulloch near Bonarbridge. A large cairn was opened at Maikel, near Bonarbridge, in 1853. In it there was a cist containing an urn with burnt human remains and a bronze blade. There are many hut circles and small mounds in various parts of the county, which may mark the site of dwellings and burial places. At the west end of Loch Migdale there is an artificial island, on which, as late as the seventeenth century, stood a dwelling, referred to by Sir Robert Gordon, who also states that a hunting seat belonging to the Earl of Sutherland was built upon an island at the south end of Loch Brora. These artificial islands may date from the Bronze Age. There is a similar artificial island in Loch Doulay. In the neighbourhood of these islands fine examples of stone implements and other relics have been found. Near Tulloch, for example,

I

there was dug up a stone cup, having handles like the wooden quaich of a later time. Some years ago a hoard of ornaments of the Bronze Period was discovered at Badbeith on the north side of Loch Migdale. A cist in a small circular cairn on Embo links was ex-

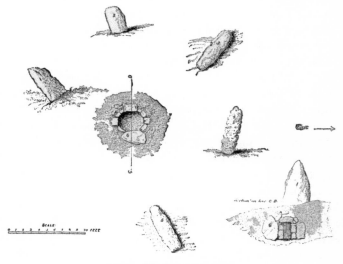

Stone Circle, Aberscross

cavated by Dr Joass in 1872. The covering slab of this cist has a cup mark which antiquarians connect with sun-worship. To the south-west of Loch-An-Trill, near Dornoch, is a cairn which appears to have contained a cist. Burial cairns of the Bronze Age occur at Aberscross, where there are other remains of this period, notably a stone circle close to the roadside, about half a mile from the Mound station.

The brochs form the chief monuments of the Iron Age, which succeeded the Bronze Period. There are one or two fairly well-preserved brochs in the county, such as Dun Dornadilla in Strathmore and Castle Cole in Strathbrora. The latter is the smallest but most strongly situated of all the Sutherland brochs. Carn

Dun Dornadilla   Doorway of Dun Dornadilla

Liath, a mile to the east of Dunrobin Castle, and close to the railway, was excavated and carefully examined by Dr Joass, who found in it shale rings, two steatite cups and two plates of brass (all now preserved in Dunrobin Museum). Brochs have been excavated at Carrol on the west side of Loch Brora, at Kintradwell in the parish of Loth (where upwards of fifty quern stones and many spindle whorls were found), at Backies near Golspie and at Sallachaidh near Loch Shin. The official report upon the historical monuments of Sutherland contains a list of sixty-seven brochs. The people of the Iron Age have left few traces in the west of Sutherland, yet the broch at Clachtoll near Stoer is one of the

Castle Cole

finest in the county. Brochs are most numerous in the parishes of Kildonan and Farr.

Hut circles, which form an important class of ancient relics in the county, are to be found all over the eastern parishes. These ancient monuments are referable to different periods, from early prehistoric times down to the comparatively recent date when the erection of rectangular buildings became common. Striking examples of these remains or circular structures occur at Swordale in Creich, Cambusmore Hill in the parish

Hut Circle, Ascaig

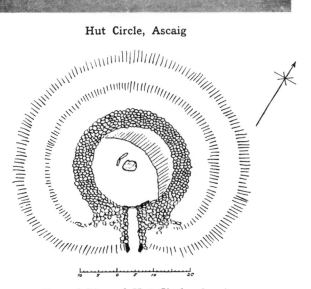

Ground Plan of Hut Circle, Ascaig

The Farr Stone

of Dornoch, Aberscross in Golspie, Achnagarron in Rogart, and Ascaig in Kildonan. An elaborate hut circle in Rogart may be said to represent a homestead of the prehistoric period.

Early Christian monuments are represented by stones carved with crosses, Christian symbols and late Celtic ornamentation. Several examples of these excised stones were found in the parishes of Golspie and Clyne when the Sutherland railway was under construction, and they are now preserved in the Dunrobin Museum. The finest example in the county of early Christian stone carving is a Celtic cross slab in the churchyard of Farr. It is $7\frac{1}{2}$ feet in height, with ornamental spirals and characteristic interlaced work, the ornamentation including a figure of two birds with their necks intertwined.

Diagram of the Farr Stone

The Farr cross is one of the finest examples of Celtic art that has come down from the early Christian period.

17. Architecture—(*a*) Ecclesiastical

There are throughout the county numerous sites of chapels which had been consecrated in the days of the early Celtic church. St Ninian's and St John's at Helmsdale, St Cardine's (or St Andrew's) at Kirkton (Golspie), St Callan's at Kinnauld and St Columba's on Eilean Co'omb, near Tongue, and at Kilcalmkil in Strathbrora are examples. There were also chapels at Gartymore and numerous other localities though the dedications have now been forgotten. Of none of these, however, do any structural parts remain. A burial vault at Inchnadamph, said to be a fragment of the old church of Assynt, is the only relic of a pre-Reformation rural church in the county.

The chief ecclesiastical structure in Sutherland is the cathedral at Dornoch. It was erected by Bishop Gilbert (1223–45) and consisted of nave, choir, and chancel running from west to east, crossed by north and south transepts, and with a square tower rising over the crossing. The nave possessed a fine window with five lights, above the west doorway, a feature which was retained at the " restoration " in 1835–37. As then restored the cathedral is 126 feet in length along nave and chancel and 92 feet along the transepts. The breadth of the nave is 25 feet and the height 45 feet.

The tower is 30 feet each way. The height of the present steeple is 120 feet.

In its original form, complete in all its parts, the cathedral must have been a graceful and attractive structure. Internally the view from the west entrance is still imposing and discloses the graceful sweep of the pointed arches. The aisles have entirely disappeared. In the seventeenth century the chapter-house, which stood to the east of the north transept, was used by the municipal authorities as the town house of the burgh till it became ruinous in 1730. In 1570 the cathedral was burnt by Sinclair, Master of Caithness, and D. Mackay of Farr, who ravaged Dornoch during a feud with the Murrays. The central tower was the only part of the building to escape destruction on this occasion. According to Sir Robert Gordon, the pillars of the roofless nave were blown over the walls by a storm at the time of the Gunpowder Plot in 1605. In 1616 Sir Robert Gordon (then tutor of the Earl of Sutherland) and other heritors of Dornoch effected some restoration of a portion of the building so as to make it sufficient for the purposes of a parish church. The choir and transepts were partitioned from the nave, roofed with flagstones from a neighbouring quarry, and provided with a gallery. As thus patched up, the cathedral served as a parish church for over two centuries until in 1835 the Duchess of Sutherland effected a more extensive restoration. Unfortunately the work was undertaken when the standard of ecclesiastical architecture in Scotland was at a low ebb. Recently the interior has been considerably improved and stained-glass windows have been added.

The church of Durness, now a picturesque ruin, was erected in 1619. An older building on the same site had been consecrated at the constitution of the cathedral, and probably a still earlier Celtic church had occupied the same site. The existing ruin consists of an oblong rectangular nave 40 feet by 16 feet, with a northern transept 25 feet by 16 feet, lighted by a side window and also possessing a gable window with two pointed lights formed by a central mullion and a transome. The gables have crowsteps and the east gable carries a belfry. The pulpit in the body of the church was placed in the centre of the side wall so as to face the transept. There is also in the middle of the church an interesting circular font measuring 1 foot 10 inches wide, 1 foot 5½ inches in the interior and 9 inches in depth. This font is one of the very few pre-Reformation relics in the county. In a recess in the south wall is the tomb of a scion of the house of Mackay who earned for himself a striking epitaph. With the usual skull and cross-bones and other insignia upon the tomb there is the following inscription :—

> Donald Makmurchov hier lyis lo
> Vas il to his freind, var to his fo
> Trve to his maister in veird and vo.   1623.

The belfry tower at Clynekirkton was used until 1825, when a belfry was added to the parish church. This tower is 11 feet 5 inches in height, 5 feet 3 inches in its interior diameter, and possesses walls 2 feet in

Tomb in the Old Church of Durness

thickness. The tower probably belongs to the seventeenth century and may be compared with those at Daviot, Ardclach, and Canisbay.

18. Architecture—(b) Military

Sir Robert Gordon writes of the castles and peels in his time (1630) thus :—

" The castles and pyles of Southerland ar Dornogh, Dunrobin (the Erle of Southerland his speciall residence), a house weill seated upon a mote hard by the sea, with fair orchards, etc. The castles and pyles of Skelbo, Pronsie, Skibo (wher ther is a fair orchard, in which ther be excellent cherries), Pulrossie, Einwershin, Golspitour, Golspikirktoun, Helmsdell (which was re-edifeid and repaired the yeir of God 1616, by Sir Alexander Gordon of Navidell, brother of John Erle of Southerland, last deceased) ; Torrish (built also by the said Sir Alexander Gordon, the yeir of God 1621), Crakok, Cuttle, Clyn, Enbo, Castle-Negoir, Durnies, Doun-Creigh, Abirscors, Ospisdale, Kentredwale, Borve, and Toung. These two last are in Strathnaver. Doun Creigh wes built with a strange kynd of morter, by one Paull Macktyre. This I doe take to be a kynd of vre ; howsoever, this is most certaine, that ther hath not been seen ane harder kynd of morter."

Of the " pyles " mentioned by Sir Robert all that now remain in occupation are the older portions of Dunrobin Castle and the tower of the bishop's palace at Dornoch. Helmsdale Castle, built in 1488, is now a picturesque ruin. A small portion of the old castle of Lord Duffus at Skelbo still stands, but the castles at Proncy, Invershin, Torrish, and Borve, are reduced to green mounds. Castle " na Coire " at Rosehall was a place of great strength, with walls seven feet thick, but it is now so much ruined that the details of the structure cannot be traced. The other " pyles " mentioned by Sir

Helmsdale Castle

Robert are now entirely gone. No part of the old castle of Skibo remains.

Of Castle Varrich, an old peel of the Mackays on a promontory near the head of the Kyle of Tongue, only the outer walls remain. They are from four to five feet in thickness. The great chimney of the old bishop's palace at Dornoch survives to show that the higher Roman clergy must have lived well. The windows of the tower may have been enlarged after the palace was burnt by the Mackays in 1570. The windows of the topmost story have not been altered. The stronghold of the Macleods of Assynt at Ardvreck on the north side of Loch Assynt was a rectangular keep. The walls remain and also a ruined staircase tower at one angle, the turret stair to the upper rooms being carried on corbelling. The rooms in the basement and in the first floor were vaulted. This castle has stood un-occupied for over two hundred years and is now in a ruinous condition.

19. Architecture—(c) Municipal and Domestic

It is significant that for many years the only municipal building in the county was the tolbooth or prison. There appears to have been a tolbooth in Dornoch prior to 1628, and it may have been used under the jurisdiction conferred upon the sheriff by the Scottish statute of 1503. After the Reformation the dignified canons had fallen upon evil days and the cathedral chapter-house, no longer needed for its original purpose, became the tolbooth. In 1730 this tolbooth had to be renovated and the new building served as council-house and tolbooth until 1813. This council-house appears to have served also during the eighteenth century as court-house for the sheriff and magistrates of the county. In 1813 Dornoch Castle was repaired and adapted to serve the purposes of a county and municipal building, while the castle tower was converted into a jail.

A handsome prison in Scottish baronial style erected in 1843 cost £3000. On the passing of the Prisons Act (1877) its use as a jail was discontinued and shortly afterwards it became the headquarters of the local

Creich House

Territorial Force. The existing county buildings, completed in 1850, form a handsome edifice with lancet windows and a piazza, where good accommodation is provided for the county officials.

There are few domestic dwellings in Sutherland older than the nineteenth century, and no dwelling older than the eighteenth century except the castles of Dunrobin and Dornoch. Dwelling houses dating from the eighteenth century are the mansion houses of Tongue, Creich, Embo, Ospisdale, and Balnakil, and one or two houses in Dornoch. Embo House is a good example of the domestic architecture of the later eighteenth century.

The shooting lodges, the larger farm houses, the

Embo House

hotels, and even the manses of the clergy have all been erected within the past hundred years. Within that time also all the houses of the smaller tenants as they now stand, have been built. Some of the shooting lodges, such as those at Loch More, Loch Stack, Lairg, Syre, Tressady, Borrobol, are houses of imposing appearance, not only comfortable but even luxurious in all their appointments. Mr Andrew Carnegie erected a massive and well-equipped castle at Skibo, which drew the admiration of Edward VII when he visited the great American in 1903.

Dunrobin Castle, enlarged and renovated in 1846, is one of the finest residences in the British Isles. It

Skibo Castle

suffered from a great fire in 1915 but is again being restored to its original splendour.

In Rogart and Creich granite is largely used for building purposes. Old Red Sandstone is used on the east coast and provides a substantial building stone which is easily worked. The Brora oolitic sandstone gives a brighter appearance than the Old Red and is also durable. Altogether housing conditions in the

K

county are good and are a marked advance upon those of former days. The " black " cottages almost universal a century ago as residences not only of smaller tenants but of farmers and middle-class gentry, have now almost totally disappeared. The thatch roofs, so pre-

Dunrobin Castle

valent in the middle of last century, yielded warmth and comfort to the cottages of that time, but have almost universally given place to roofs of iron and slate. Though an advance from the economic stand-point, these modern roofs lack the picturesque beauty of the older style. The older dwellings of the county were gravely defective in ventilation and drainage. The advance in sanitation within the past half century has been not less striking than the improvement in architecture.

20. Communications—Past and Present

Prior to 1807 there were no roads in Sutherland and no bridges except Brora Bridge. From the Meikle Ferry to the Ord a horse track ran along the sea-shore, and by similar tracks Strathnaver and Assynt communicated with the ferries at Portinlick and Bonar, but no wheeled vehicles were in use in any part of the county. The access to Sutherland from the south of Scotland was mainly by sea. On the mainland the county was reached by the Meikle Ferry (G. *Port a' Choltair*), Bonar Ferry (G. *Am Bhannath*) and Portinlick (G. *Port na-Lice*). At these ferries the droves of cattle for Falkirk trysts and other southern markets swam the Kyle in their southern journey. It was currently believed among the people in these old days that if an ox readily took to the water and crossed the ferry without turning back it would be sure to fetch a good price at the market.

The Sutherland portion of the road from Inverness to Thurso (which traverses the east coast from Bonar by Evelix, and the Mound to the Ord) was completed in 1812–13, when the Kyle was spanned by the Bonar-

The Mound

bridge. Within twenty years from that date upwards of 400 miles of roads were constructed throughout the county. Roads were formed from Bonar by Lairg to Scourie and Durness, as well as to Tongue and Strathnaver. The road to Lochinver by Oykell Bridge and Inchnadamph brought Assynt into touch with the east of Scotland while that from Helmsdale by Forsinard to Melvich, as well as the north coast road from Melvich by Bettyhill and Tongue to Durness, connected the Reay country with easter Sutherland and Caithness. Useful roads were also made about the same time from Bonar by Loch Buidhe to the Mound, from Lairg to Rogart and from Rogart by Sciberscross to Brora. Scourie was connected with Assynt by the road to Kylescu Ferry, which was continued so as to join the Lochinver road at Skiag Bridge. Though these roads were adequate for the traffic in the county at the time they were constructed, they had not sufficient depth or strength of crust to bear heavy traffic. They were, however, a great boon to the people of the county. In 1819 a mail-coach began to run from Tain to Thurso with a daily service of mails, and from that time regular daily intercourse with the south of Scotland gradually extended to the remoter parts of the county.

The advent of motor traffic towards the end of last century soon put a great strain upon the Sutherland roads. Between 1900 and 1915 the outlay upon these roads increased fourfold and now amounts to about £12,000 per annum, or one-eighth of the whole rental of Sutherland. Since the creation of the national

Road Board in 1910, advances have been received by the county from the Board for strengthening and draining the roads and for improving the crust so as to bear the greatly-increased traffic. It cannot, however, be said that the road problem in Sutherland has yet been solved. The upkeep of fully 500 miles of roads with much extraneous traffic is too great a burden to impose upon the local resources, even with such help as is now given from central funds.

The Sutherland Railway from Ardgay to Helmsdale was constructed by the third Duke of Sutherland in 1870, and was extended in 1872 (as the Sutherland and Caithness Railway) by Kildonan and Forsinard into Caithness. These railways were afterwards absorbed into the system of the Highland Railway Company. The small gauge railway from Dornoch to the Mound was constructed in 1896 and, though worked by the Highland Railway Company, has continued its career as a separate concern. While the eastern part of the county thus acquired improved access to the south of Scotland, the northern and western portions continued for many years to depend upon horse traction of mails and passengers until the introduction of motor power early in the present century. There is now a motor service from Lairg station to Lochinver, to Scourie, and to Tongue, while the villages on the north coast are in regular motor communication with Thurso.

21. Administrative Divisions

With two exceptions—Tongue and Eddrachillis, disjoined from Durness in 1726—the parishes of Sutherland, which are the oldest administrative units, date from the second quarter of the thirteenth century. Until the beginning of the sixteenth century the territorial administration was exclusively ecclesiastical. The judicial authority of the barons was limited to their own lands, and the limits of the jurisdiction of the sheriff courts established at Dornoch in 1503 remained undefined. The right of " regalaty, sheriffship, and crownerie " of the new shire of Sutherland was made over in 1631 to the Earl of Sutherland, in whose family it remained until the abolition of heritable jurisdictions in 1747. The new shire was represented in the Scots Parliament in 1639 by Robert Murray of Spinningdale, whose brother, Walter Murray of Pitgrudie, sat for the burgh of Dornoch in the same year.

After the outbreak of the Civil War the authorities in 1643 found it necessary to constitute Commissioners in each shire for providing supplies for the Scots army. These Commissioners of Supply soon proved useful county authorities for general purposes and continued to fulfil a variety of public duties until the introduction of representative government in 1889. Since that date county administration (including the control of roads, public health, licensing, and police) is under the charge of a popularly elected County Council and various

County Committees. For road and public health administration, Sutherland, like several of the smaller counties, forms a single district.

The Court of Lieutenancy of the Lord-Lieutenant was established on its modern basis by the provisions of the Militia Act of 1802. The military powers of the Lord Lieutenant were transferred to the crown in 1871 but restored in 1907 when the territorial forces were reorganized. The Sutherland Territorial Association administers in peace time the local squadron of the Lovat Scouts Yeomanry and the Fifth Battalion of the Seaforth Highlanders.

Boards for the administration of poor relief were constituted in each of the thirteen parishes in the county by the Poor Law Act of 1845. After a useful career of half a century these parochial boards were, in 1894, replaced by elective parish councils. The county combination poorhouse at Bonarbridge, provided in 1870, was never a popular institution, as the people prefer the freedom of cottage life. The introduction of old age pensions in 1908 was followed by a fall in the number of persons in need of parochial relief.

Prior to 1872 there were parish schools and one or more Free Church schools in every parish. After the constitution of School Boards under the Education Act improved schools were erected throughout the county and there are now fifty-six efficient public schools in Sutherland, while Higher Grade Schools have been provided at Dornoch, Golspie, Brora, Helmsdale, Lairg, and Durine. The Golspie Technical School, under private management, is a residential institution endowed with a number of bursaries. The county Education Authority provides bursaries for higher education leading to the universities and to central institutions. There is an advisory committee in connection with the work of the North of Scotland College of Agriculture.

22. Roll of Honour

The earliest notable man associated with the county was Sigurd (d. 875) who gained permanent fame by his invasion of Scotland. He was buried on the top of a hill near Dornoch. In the Roman Catholic period the greatest name is that of Bishop Gilbert of Moray (d. 1st April 1245). An ecclesiastical statesman of good standing, St Gilbert (the last Saint in the Kalendar) held important administrative offices. His literary work (in Latin) included *Exhortationes ad Ecclesiam Suam* and an essay *De libertate Scotiæ*. Sir Robert Gordon (1580–1656), notable as a diplomatist, administrator and man of letters, wrote a *Genealogical History of the Earls of Sutherland* (published in 1813) containing much information upon Scottish history. In the Covenanting struggles of the seventeenth century the Earl of Sutherland took a prominent part and was the first to sign the National Covenant in 1638. Donald Mackay of Farr (who in 1628 became the first Lord Reay) distinguished himself in the German religious wars under Gustavus Adolphus of Sweden. General Hugh Mackay of Scourie (1640–92), the opponent of Claverhouse at Killiecrankie, rendered important military service in Ireland and the Netherlands. He wrote *Rules of*

John, Thirteenth Earl of Sutherland

War for the Infantry. His *Memoirs* by John Mackay of Rockfield, a native of Lairg, appeared in 1836. The most notable name associated with Sutherland in the

eighteenth century was George Dempster of Skibo
(1732–1818), next to Sir John Sinclair, the greatest con-

Lieutenant General Hugh Mackay

temporary follower of Adam Smith. He was for nearly
thirty years M.P. for the Forfar burghs, and wrote
several works upon economic and political topics. Sir

James Matheson (1796–1878), a native of Lairg, and the third Duke of Sutherland (1829–92) both spent large sums upon land improvement and railway extension. The *Memorabilia* of the Rev. Donald Sage (1789–1869) secures for him a place among notable Scottish writers. Few books shed a shrewder light upon the social condition of the north of Scotland in the eighteenth and early nineteenth centuries. The Rev. Mackintosh Mackay (1793–1873), Editor of the Highland Society's *Gaelic Dictionary*, and of Rob Donn's *Poems*, a prolific Gaelic author, was a native of Eddra-chillis. Alexander Munro (1825–71) a Sutherland man by descent, was a sculptor of good repute, who for many years exhibited at the Royal Academy. The Rev. J. M. Joass, LL.D. (1838–1914), for over forty years minister of Golspie, was an eminent archaeologist. The Rev. Angus Mackay, Westerdale (1860–1911), a native of Strathy, occupies a worthy position among the historians of the north of Scotland. He wrote the *Book of Mackay* (1906) and the *History of the Province of Cat* (1914).

Among notable sons of natives of Sutherland were Sir John A. Macdonald (1815–91), the great Canadian statesman ; Dr George Matheson (1842–1906), author of many books on philosophical and devotional subjects ; Dr Robert Rainy (1826–1906), a notable ecclesiastical statesman, and Dr Charles Mackay (1814–89), the well-known poet, whose father, Lieut. George Mackay, belonged to a Strathnaver family.

In Gaelic literature Sutherland produced one poet

in the front rank and several lesser lights. Bishop
Gilbert's Gaelic version of the Psalms is no longer
extant. Gaelic hymns by John Mackay, Mudale,
became popular in the seventeenth century as a means
of conveying religious instruction when no Gaelic
version of the Scriptures existed. Between 1750 and
1850 a succession of Gaelic poets flourished in Kildonan,
of whom the best known were Donald Matheson, Badanloch (1719–82) ; and his son, Samuel Matheson (d. 1829).
Donald Matheson was a contemporary of Rob Donn,
who claimed to have " more poetry but less piety "
than the Kildonan bard. Lieut. Joseph Mackay, a
Waterloo veteran, wrote Gaelic *Laments* upon eminent
people of the Reay country. Robert Macdonald (1795-
1870), a native of Loth, wrote Gaelic hymns.

Rob Donn (1714–78), the great Gaelic bard of the
northern Highlands, like many Gaelic poets of his time,
was quite illiterate. A poet of nature in the truest
sense, he excelled in realistic description and in the
expression of emotional feeling. His love songs and
elegies sank into the hearts of the people and became
widely known by oral transmission before any of his
work had been committed to writing. His satire upon
Robert Gray of Creich is a good specimen of his pungent
style. Two brothers, George Ross Gordon and William
Gordon, soldiers, one in the Reay Fencibles, and the
other in the Black Watch, who published Gaelic poems
in Glasgow in 1804 and 1819, the worthy Dr Thomas
Ross of Lochbroom (1760–1845), author of a Gaelic
Psalter and editor of the second edition of the Gaelic

Bible (1807) and John Munro (1791–1837), author of
the song *O, théid sinn, théid sinn*, whose poems were
published in Glasgow in 1819, were all natives of Spinningdale. Donald Mackenzie (1768–1861), for sixty

Dr Gustavus Aird

years catechist in Assynt, his native parish, maintained
in his *Laoidhean Spioradail* the tradition of the Kildonan
poets.

Dr Gustavus Aird of Creich (1813–98), a notable
Gaelic preacher, was an authority on the ecclesiastical

traditions of the northern Highlands.   The Rev. George Henderson (1872–1913), minister of Eddrachillis, besides editing the poems of John Morrison, Rodel, and the *Memoirs* of Evander MacIver, Scourie, published volumes upon *The Survivals of Belief among the Celts* and *The Norse Influence on Celtic Scotland.*

Lord Kennedy (Neil John Downie Kennedy, 1854–1918), the first Chairman of the Scottish Land Court, who made notable contributions to Scottish jurisprudence, particularly in regard to land rights, was born at Rosehall.

Andrew Carnegie (1837-1919), during the last twenty years of his life, spent the summer and autumn at Skibo, his Scottish residence.

23. Chief Towns and Villages of Sutherland

(The figures in brackets after the names give the population in 1911, and those at the end of each section are references to pages in the text.)

N. = Norse.   G. = Gaelic

Bettyhill (named after Elizabeth, Countess of Sutherland), is a centre of agriculture and fishing.   (p. 149.)

Bonarbridge (392), G. *Bhannath*, " the low ford," founded in 1817 where previously there had been a ferry and ford for cattle.   Bonar district is noted for fine scenery and possesses many antiquarian remains.   (pp. 87, 92, 103, 118, 129, 147, 149, 152.)

Brora (572), N. *Brúar-á*, " bridge-river," established at the beginning of the seventeenth century as a burgh of barony, and revived a century ago, is now the chief industrial centre in the county. It possesses a wool mill, distillery, brick works, coal mine, quarries, and marine and river fishings. It is also a summer resort, and possesses a good golf course. (pp. 94, 112, 115, 118, 119, 131, 136, 138, 145, 147, 149, 152.)

The Witch's Stone, Dornoch

Dornoch (741), G. *Dorn-achadh*, " pebble-field," is the county town, a notable watering-place and fashionable summer resort. St Bar's Church, founded in the eighth century, is no longer extant. A great part of the bishop's palace (probably begun in the thirteenth century, and extended in the fifteenth and early sixteenth centuries) was burnt in 1570. The tower with corner turrets and crow-step gables is still occupied. The Academy was erected in 1912. The Dornoch Grammar School was endowed in 1641, and the burgh had the services of several successful teachers in bygone times.   The burgh received its charter in 1629, and its records are complete from 1731.   Possessing two excellent golf courses, Dornoch is one of the best known golfing resorts in Scotland.   The last execution for witch-craft in Scotland took place at Dornoch in 1722.   (pp. 80, 81, 85, 102, 103, 104, 115, 122, 123, 124, 125, 130, 135, 136, 137, 140, 141, 142, 144, 150, 151, 152, 153.)

Smoo Cave, Durness

Durness, N. *Dyranes*, " wild-animal headland," at one time the hunting-seat of the bishops of Caithness, and later of the Lords of Reay, possesses fine surroundings.   A mile east of Durness is the cave of Smoo, 110 feet wide and 53 feet high at its mouth, and penetrating into the limestone a distance of 450 feet.   (pp. 88, 92, 93, 99, 102, 103, 115, 138, 140, 149, 152.)

L

Embo (588), (c. 1230), *Ethenboll*, in Gaelic *Eireabol*, was founded about 1820 and is the most populous fishing community in the county. Situated at the innermost shore of the Moray Firth, Embo adjoins valuable fishings, but greatly needs improved harbour accommodation. Its inhabitants, mainly Celtic, retain some of the older Celtic customs. (pp. 103, 104, 116, 120, 130, 140, 144.)

Golspie (1046), N. *Gols-by*, the most populous village in the county, is in a richly wooded country. Near it is Beinn-a-Bhragie, on which is a monument to the first Duke of Sutherland. Dunrobin Museum contains a representative collection of the antiquities of the north of Scotland. (pp. 74, 104, 116, 119, 128, 131, 135, 136, 140, 152, 156.)

Helmsdale (752), N. *Hjalmunds-dalr*, with the best harbour on the east coast of Sutherland, was long an important centre of the fishing industry. (pp. 92, 93, 103, 104, 116, 118, 119, 127, 136, 140, 149, 150, 152.)

Kinlochbervie, G. *Cean-loch*, "head of loch," N. *Bergje* "rocky," on the north shore of Loch Inchard, with striking rock and headland scenery. (p. 100.)

Lairg (336), G. *Lairg*, "the pass," a growing village, is the centre for Lochinver, Scourie, Durness, and Tongue. There is here a handsome memorial to the late Sir James Matheson. The antiquarian monuments of the district are numerous. (pp. 82, 88, 108, 112, 113, 115, 128, 144, 149, 150, 152, 154, 156.)

Lochinver (165), G. *Loch-an-inbhir*, "river mouth," possesses a good harbour, and is the centre of magnificent mountain scenery and excellent fishings. (pp. 89, 100, 115, 116, 118, 119, 149, 150.)

Melvich, N. *Mel-vik*, " sand-bay," and **Portskerra**, G. *Port*, " harbour," N. *Sker-je*, " skerry," industrious fishing villages.   (pp. 92, 116, 149.)

Rogart (c. 150), G. *Raord*, situated in romantic surroundings, is growing in popularity as a health resort.   (pp. 108, 112, 115, 135, 145, 149.)

Scourie (77), N. *Sker-je*, " skerry," the centre of extensive loch fishings.   (pp. 89, 100, 116, 149, 150.)

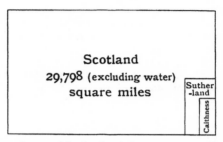

Fig. 1.   Areas of Sutherland (2028 square miles) and
Caithness (686 square miles) compared with
that of Scotland

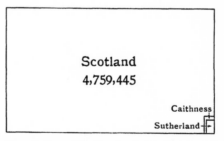

Fig. 2.   Population of Sutherland (20,180) and Caithness
(32,008) compared with that of Scotland in 1911

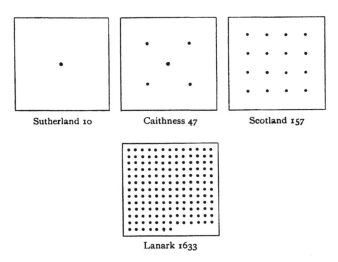

Sutherland 10   Caithness 47   Scotland 157

Lanark 1633

**Fig. 3. Comparative density of Population
to the square mile in 1911**

(Each dot represents 10 persons)

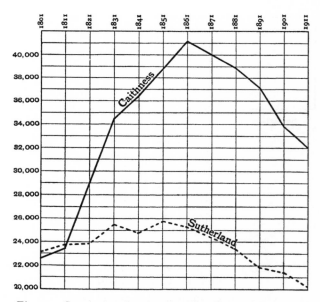

Fig. 4.   Graph showing comparative rise and decline of
population in Caithness and Sutherland during
the last 100 years

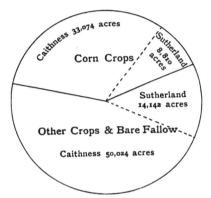

Fig. 5. Proportionate area under Corn Crops compared
with that of other cultivated land in Caithness
and Sutherland in 1917

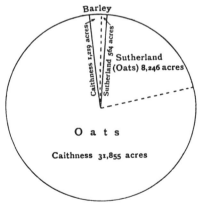

Fig. 6. Proportionate areas of Chief Cereals in
Caithness and Sutherland in 1917

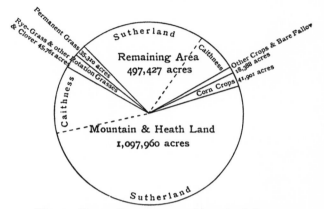

Fig. 7.   Proportionate areas of land in Caithness
and Sutherland in 1917

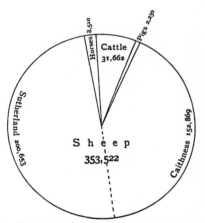

Fig. 8.   Proportionate numbers of Live Stock in
Caithness and Sutherland in 1917

www.ingramcontent.com/pod-product-compliance
Ingram Content Group UK Ltd.
Pitfield, Milton Keynes, MK11 3LW, UK
UKHW042144280225
455719UK00001B/83